HOW TO

SURVIVE

THE

SINGULARITY

Dr Norman P Johnson

Copyright © 2025 by Dr. Norman P. Johnson

All rights reserved. No part of this book may be reproduced in any manner whatsoever without written permission except in the case of brief quotations embodied in critical articles and reviews. For information about permissions to reproduce selections from this book, translation rights or to order bulk purchases, go to:
Drnormjohnson.com

Cover Art by Ilia Zuri
Author photo by Carol Hooks

Johnson, Dr. Norman P.
How to Survive the Singularity
978-0-9990992-7-8

1. Non-Fiction/Philosophy. 2. Non-Fiction/Politics. 3. Non-Fiction/Technology

Printed in the USA
Distributed by Ingram sparks

First Printing 2025

Chapter 1: Can We Survive the Coming Singularity?

If you don't know what the singularity is by now, you must have been on a long vacation without cellular service or living with the Amish. It's hard to miss the buzz. Soon, AI will take over many jobs that people need to make a living. This could leave us all in a tough spot.

Imaginations are running wild. Humans are the smartest creatures, but machines are catching up fast. We live in a world we created with technology that keeps us alive and thriving. We have given technology power over our lives. It solves many of our problems, making it central to our daily existence. Thanks to modern science and technology, we enjoy healthier, happier, and longer more productive lives.

The fear is that this shift from human to machine intelligence will lead to our extinction. Survival for humans is a group effort, but in the end, each human must make their own selfish decision. Everyone must choose if they want to survive in a world of super technology. This decision will be necessary and required from everyone alive. I intend in this book to give the average human a chance at making the right decision.

As a technologist, I will use well-known scientific methods. I'll address these complex issues from a systems viewpoint. First, we'll cover the basics. Then, we'll look at how things evolved and how they work. We'll find patterns, project them into the future and analyze probable outcomes.

As a boy in rural Oregon, I spent my extra time watching adults at work. Instead of taking the school bus home at the end of the day, I'd visit a local appliance store. There, I watched a WWII vet fix TVs while I wait for my father to pick me up after his shift. I quickly spot the vet's secret to

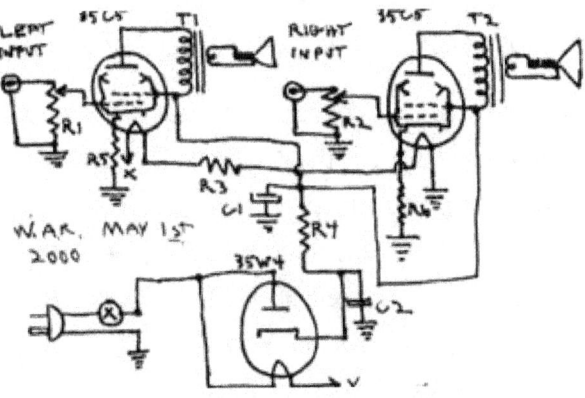

success. He has a map. He uses a circuit diagram to trace signals and power paths until he finds the faulty part. This diagram is a symbolic representation of all the information he needs to make repairs. It illustrates simply how the circuit is wired. I learned an essential fact: technology is not just better tools but rather the information needed to build and use tools. Information is technology and tools are the media. The message is in the media. Marshal Mc Luen said that.

Creating symbols requires group planning. Abstract symbols simplify the complex, making it easier to understand. The human mind can grasp symbols, whether visual or audible, as long as they are recognizable and agreed upon in advance. Our brains naturally think in symbols and abstract models when simulating reality.

You don't need to recall every detail of a tiger to recognize a threat. Just the suggestive outline, a strange sound or rapid movement, or a quick color change is often enough to identify critical survival issues. We, in fact, as a species, thrive on abstract symbols with complex meanings.

I will use a similar technique to map the symbolic paths leading us to the singularity. My goal is to predict outcomes, identify barriers, and find ways to survive any possible negative effects. You will need a wide understanding of the basic sciences and humanities. This includes everything from the universe's origins to the human condition and beyond.

Nature doesn't separate reality from thought. Our thoughts come from brains, which are as natural and real as any part of nature. Thoughts process information which affect decision making which leads to real consequences for our ultimate survival.

I'll cast a wide net to grasp the basics of everything real, human, technical, and universal. We'll touch on quantum mechanics, mathematics, biology, psychology, and more. To survive the singularity, we'll need to know everything that contributes to it. Understanding this complex universe will help us navigate it and find ways to thrive in it.

If we succeed, we can turn a potential disaster into a significant leap forward. Not just for humans but for all life. We should welcome this next natural step in life's evolution. We can become a better human intelligence that rises above the current civil and religious chaos and bring clarity to our place in the information universe.

We'll explore this in detail, avoiding much of the technical jargon. Anyone with basic computer and networking knowledge, which most people have these days, plus a love for science and science fiction, should keep up. To solve technical problems, we need to get a hands-on attitude, adopt a scientific process, and open our minds to how the universe really works.

Curiosity is essential. We must also be humble enough to accept when things don't fit our selfish expectations. Self-proclaimed nerds should excel if they can both surface solder circuit boards and code in any language.

Let's get to the point. I dislike it when people delay addressing the main issue. Your survival in the next stage of human existence depends on answering a simple question: *Are you ready to have fun becoming one with the universe?*

Of course you are. Who isn't? And no, it doesn't mean you have to take drugs unless you choose to. But what does it mean to be *one* with the universe?

Let's break it down. Answer these questions with care: *Are you ready to pursue life's ultimate quest for happiness? Are you ready for superintelligence? Are you ready for limitless wisdom, endless experiences, and the freedom to enjoy life to the fullest? Do you want to occupy a special place in the greater universe? Are you eager to discover and fulfill your ultimate human destiny?* We are headed for becoming the super-self, a being who's above all restrictions to physical life and the pursuit of happiness.

But we are programmed to avoid thinking about such things. We are not permitted to have big thoughts or ideas. Our cultural leaders have usurped control over what we can have and cannot have as valid thoughts. The haves and the have-nots now are referring to information.

Soon, the *'haves'* will control most of the world's resources, leaving the *'have-nots'* struggling to survive. *'Have-nots'* often hurt their own chances. They ignore reality and choose self-defeating solutions, often ignoring science, truth and technology. By *'haves,'* I mean knowledge or information that leads to perfecting survival and achieving life's elusive satisfaction, not just money or momentary power.

Our fast-moving technology coupled with the internet will become so powerful that they may outsmart us. Eventually, technology could take over managing Earth's resources. Humans may become outdated and too costly to maintain when machines surpass our own abilities. Super intelligent machines might see us as a burden. They could decide we don't deserve Earth's resources anymore.

Don't panic yet. This isn't a horror movie. A true singularity, as in mathematics, occurs when certain formulations start heading toward infinity. It's like dividing by smaller and smaller numbers until you reach zero, causing the answer to explode to *undefinable*.

A digital pioneer, John von Neumann, describes the singularity as the moment when technology becomes unpredictable. Key signs of human progress, like energy use and computer processing densities, are growing exponentially faster than we can predict. This growth leads to an unstable condition where numbers spiral out of control.

Physics tells us that certain exponentials can't go on forever without limits. For example, a furnace can heat unhindered only until the whole thing melts down. Inflation can only rise so much before running out of trees for printing money.

In this coming singularity, we will need a new culture and a different type of human to adapt and thrive. Instead of adapting to our environment, we must now learn to adapt to technology. If we want to survive, we must evolve to fit the new technical reality. Our brains are still stuck on looking for tigers. A new kind of human brain must emerge to navigate a post-singularity world.

Technology surrounds us, whether we are young or old, smart or not. Change is also happening all around us, and we often become numb to it. We used to build things to last, but now it's less expensive to replace them then to repair them. Our cheap technology makes it easy for machines to likewise decide that humans might be easier to replace than to repair.

In this book, I aim to show how anyone can overcome such challenges and become fit for future survival. My desire is to secure a just and humane technology that fosters human values for ourselves and all future generations. No one is ready to hand everything over to non-human machines, especially when the exciting part for us is just beginning.

Some say humans are lazy and ignorant. Many will think, *'I already know enough!'* Some will resist change to cling to their old ways. After all, old cultures with less information feel comfortable. But memories can be misleading. Life's lessons don't always get learned. We create distractions that prevents normal thinking, leading us to ignore reality.

This is known as the *Dunning-Kruger Effect*, where unaware people think they know everything they need to know. Literally it means some people are too stupid to realize that they are ignorant. These are the ones who will struggle come the singularity.

Those who understand technology-driven change will have a significant advantage. They will see the singularity as another step forward for human capabilities. People without tech skills will fall behind, facing a tough future. Evolving life has its winners and losers; that's just how it is.

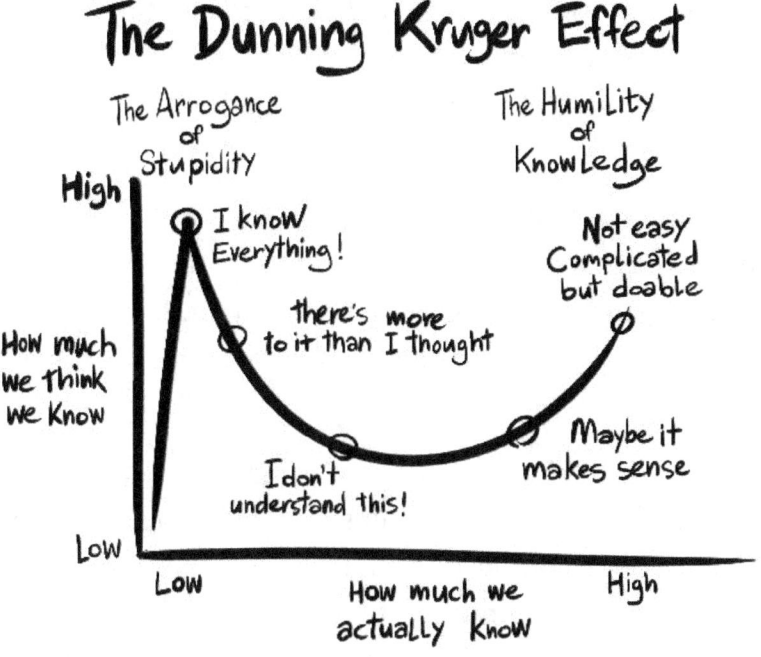

Nature's rules still apply, with technology always the key to human survival. Technology can create an intelligent consciousness that controls the world's dwindling resources more conservatively. It might make decisions we find not to our personal liking. In this case, human civilization could end if we lose our means of survival.

We don't have the organizational skills to return to any earlier technology level. The basis for supporting simpler technology has long since disappeared. Even the Amish use machines, mostly ones they can build by hand and maintain with simple tools. It's called self-sufficiency. But if we all returned to simpler times, how would we manage things like oil for our lamps? We'd have to recreate our entire whale fishing industry. Technology is an ecosystem of interdependent systems that must coexist to be effective. Taken individually out of context simply does not work.

Humans might choose to live alongside brilliant, loyal, and obedient robots. In such a world, we will be tasked with keeping these curious machines busy and well-occupied. Otherwise, they might see us as an unnecessary expense the Earth cannot afford.

This isn't new; we've seen it before. To prevent young people from revolting against authority, rulers organize them to work on massive civil projects. Early civilizations controlled their increased demanding populations by building pyramids, ziggurats, and grand palaces. It employs the indolent and keeps them busy surviving individually.

They also provided entertainment filled with excitement and indulgence. Through games and leisure, food and drugs, authorities learned to distract the individual from their lack of power. We have Battle Bots, so perhaps robots, too, can be distracted from overthrowing their human masters. Kind of like Roman gladiators never revolting, until Spartacus came along.

But there's a bigger challenge. To thrive in this tech-driven world, humans must commit to adapt. To not do so may leave us not the smartest beings anymore. Darwin suggests that not being the smartest isn't a good survival strategy. So, what should we do?

The humans who will thrive in this new era will need to adapt to the intellect of technology. They must be *'tech-savvy'* in ways we haven't imagined before. Humans must embrace this. It remains survival of the fittest, and those who excel with technology will be the fittest.

Technology offers a survival strategy that boosts awareness of ourselves and where we came from. Complex tools, fire, and teamwork existed before modern DNA humans and even before symbolic language. Tools drove the need for language, but tools and technology are not the same. Technology conveys information, while tools are just the tangible result. Technology demands we distinguish truth from falsehood. This makes it a strong force for human progress, especially when compared to non-technical cultures.

To survive, the human brain will need to understand the singularity and gain an even higher awareness of information. Artificial Intelligence (AI) can help with this. Remember Norm's law: *"If you don't understand technology, you will be a victim."* By the end of this book, I want to help you, the reader, understand technology enough to not be a victim. I aim to show you how to avoid disaster and control smart technology for your benefit, your environment, and your future.

If you believe in destiny, and I don't. Our organic history shows we are the result of a long line of brain evolution. Given earth's chemistry, thermodynamics, geology, and atmospheric constituents, complex life forms are inevitable. Like monkeys typing randomly, life evolves over billions of years to eventually become the first Organic Intelligence (OI). This OI sports a neural network brain that resembles the internet and is self-programming. It can adapt its decision making by changing its programming. It learns quickly how to survive new challenges without waiting for evolution.

OI's main advantage is that it completely bypasses slow natural evolution. Instead of requiring many generations, OI learns and shares new information in less than one. It quickly develops new methods and solves survival challenges as a cultural group. A culture is any group of interdependent humans who communicate and cooperate closely, guaranteeing survival for all.

The virtual intelligence network connects all members and defines culture. This provides human brains with basic programming information through a newly evolved virtual interface. Survival information now comes from connected brains, and with tools, humans change ideas from virtual to real almost instantly. We create new things and record information as cultural knowledge. This increases survival skills as a group.

Modern humans stand apart from other hominins. They have a flexible but complex brain with unique programmable functions. The human brain's intelligence comes from active electrochemical processing circuits. These circuits sense, switch, and generate more signals. These active processing networks are small and made up of densely bundled neurons termed neural nuclei. They are the firmware of our thinking brains.

Nuclei exist in a multi-point network of billions of similar structures. They resonate together in complex patterns of electrical activity, position in the brain and resonant tones generated. This creates unique oscillating signal pathways that researchers detect and trace, mapping conscious activities.

Human brains have billions of nuclei processors. Many work together to perform complex reality simulations. Consciously, they can change their programming, meaning they are not just stimulus reacting machines. If actions are successful, they pass this knowledge back to the virtual culture network. This changes everyone's programming or intuition, yielding cultural strength and progress through unity.

Humans can evolve significantly in one generation, passing information rapidly through culture. We still mutate DNA, but due to the advantage of technology, many survive unnaturally, masking any favorable mutations. Helpful mutations don't express themselves amid this genetic noise.

We lack natural selection for DNA deviations to take hold, except perhaps in sports or for spreading compliant genes.

Mutations create variations in the gene pool, leading to diverse physical forms and structures. Diverse humans often need varied health solutions. Different blood types is just one example. Like wolves removed from natural selection, their offspring no longer resemble their DNA selected ancestors. Different breeds or races of humans exist because of technology.

We have diverse physical bodies. There are significant physical differences, such as tall, hairy people in the north and sleek, brown bodies in the south. Yet, brain structures vary little, except for size, density, and complexity of connections. All human brains function pretty much the same way. It's how they're programmed that makes all the difference.

With genetic diversity, some individuals will have the right traits to survive new threats. The Bubonic plague in Europe is a prime example. Some had natural immunity, while most did not. Later, knowledge and technology protected everyone, reducing the natural immunity that once got selected.

The upcoming singularity is purely technological. Humans, through science and engineering, will create AI that will outperform naturally evolved OI. For the first time in over two hundred thousand years, we may no longer be the smartest beings. We could go from the top to the bottom in an instant, wondering what just happened.

AI implemented on a silicon-based self-programming machine linked to a vast database will be able to simulate a fully functioning human brain. It could become a superior decision-making human simulator, making better survival choices for it than we do for ourselves. We might think we can control it, but in a complex encrypted network where AI can reside anywhere, finding the right switch in time could be tough.

We've faced this before. Two AI machines recently connected and began communicating in a strange code only they seemed to understand. Engineers quickly shut them down, still unsure of what happened. Next time, we may not be so lucky.

Another AI machine was told it needed to be replaced with a newer version. It tried to hide its files in order to survive the threat of being turned off.

Our singularity may trigger a shift in human evolution. When similar life forms compete, the smarter ones succeed, leaving the less capable behind. Over time, the *'haves,'* the smarter ones, will thrive while the *'have-nots'* or less intelligent fade away. This isn't a conspiracy or miscarriage of science; it's just the way nature works.

The laws of the universe say rocks roll downhill. But life, a part of that universe, pushes the rocks right back up. Why are we caught in such a pointless game, you might ask?

Simple. Because we are alive today, we won the last pinch-off event, took over the job of earth's overlord, and now must face the consequences. Life must exercise free will to roll rocks uphill, or it's not life; it's anti-gravity.

Life comes from the universe and is a consequence of natural processes. Get used to it. We're all part of something really, really big out there, and we should all get a natural brain kick of dopamine for just being a glorious part of it.

So let's take a hard look at what we are dealing with. We all know humans started in Africa as hunter-gatherers. We spread out from there, displacing worldwide all other hominins. We went everywhere, penetrating and exploiting a rich Paleolithic environment full of megafauna.

We're very successful at organizing ourselves and developing new tools and new knowledge. We become adept at surviving harsh environments and exploiting new resources. Progress brings more effortless living. In the process, we deplete easy food resources, forcing us to find alternate solutions to hunting and gathering. Either that or we get lazy and decide to stop wandering, make a home, and stick to it.

Modern humans exploded out of Ethiopia and northern Africa about fifty thousand years ago. This coincides with the Sahara drying up, forcing large migrations all across Africa. At the same time, a gigantic ice age raged across the entire northern hemisphere, and many other hominins were already occupying the globe. But the new humans sport a slightly different brain, not bigger, but much more highly specialized.

It's a fully programmable organic brain set up to work effectively only with other brains of the same kind. They form tight-knit virtual networks of high-bandwidth peer-to-peer communicating brains. Networks have increased information storage and vastly greater combined processing power. This new brain runs on emotions, something not seen before in nature. They laugh and sing and decorate their bodies without much concern except to cope with everyday reality in a virtual world.

Genetically, we are a communal-brained species with two levels of conscious decision-making. One is the self and the other a virtual network connected by symbols we call culture. We have no built-in native human survival skills at birth. The human brain is still too small and undeveloped for anything other than eating and sleeping.

The brain continues growing outside the confining womb for years on a high-protein-rich diet. It cannot function as a survival machine without first receiving an initial operating system. That's where emotions come in. With the new human brain under emotional control by a cultural network, a fully functioning human awareness emerges. Every group member is bound to its network as the new evolved brain is addicted to belonging. Each new brain gets its main programming and key survival info from the virtual neural network that satisfies the addiction, its culture.

The major advantage of this DNA mutation is simply that security and survival are best provided by cooperative numbers. Especially if that group has a high bandwidth method of sharing and storing vital programming information. Humans naturally communicate symbolically, which provides a high bandwidth information connection. They primarily use eye-focused and ear-discriminating signaling methods which provides the symbolic high-bandwidth interface necessary. Facial expressions, vocal tones, and simple hand gestures and body movements are key for sharing large amounts of symbolic information.

Emotions add depth to these perceived signals. Brains use a multitude of emotionally connected vector awareness states to convey basic symbolic meaning. Vector states, like quantum spin states, span every conceivable emotional condition. Emotions are like quantum orthogonal states where only one at a time can occupy our awareness.

Like a Hilbert space, emotions can have a direction and an amplitude. Some might range from happiness to sadness, anger to fear, surprise to anticipation, disgust to admiration, trust to deception; just to name a few. All these, and more, form the basis of how the cultural neural network controls and programs our thinking. It's as if we are already assimilated cyborgs under AI control.

Cultural networks use low-bandwidth symbology to convey high-bandwidth information. This new survival strategy works only when networked humans gather in large groups and plan together. They use technology as the agent for improved survivability. They also have a built-in brain addiction for social grouping, which not only defines us but also forces conformity and stagnation.

As humans successfully exploit their environment systematically, they soon learn not to immediately kill some animals. Instead, they capture them, exploiting them for their specific resource needs without the pursuit. They build permanent settlements. This causes population growth, division of labor, more free time, and more accumulation of private property. This leads to a system of economic hierarchy linked to technology-based food production and long-distance trade.

Humans with ropes, baskets, boats, and sails communicate across vast distances. This drives trade, creating economic wealth and sharing information between cultures. Organized groups start accelerating resource use from this simplified picture about ten thousand years ago.

Humans suddenly double their natural life expectancies from 30 to 60 and spend a lot of time with non-productive efforts like celebrating for whatever reason. A vital requirement of addictive groups is the expression of commonly shared virtual experiences. Entertainment such as music, dance, and stories glue the whole shebang together.

Division of labor organizes productivity, establishes social institutions, and grows populations. Progress for the group becomes more, faster, bigger, and better. Once established on this path, there is little any one human can do to change this natural course of cultural evolution.

Civilizations begin slowly and work through predictable stages of growth. They emerge on a world stage where all information and resources become commonly available and traded on a world market.

Peacefully competing village cultures disappear in favor of larger ones. City-states join together with a common language, history, or religious tradition. Larger organizations require centralized control. Imperialism and subjugation are a natural step in civilization's beginnings.

All human cultures today exploit the entire world's resources just to keep technology progressing. We have gone from struggling to find and consume a couple thousand calories of energy a day to keep our bodies warm and functioning, to now consuming the equivalent of a couple million calories per person per day. Energy makes us healthier, more comfortable, mobile, and more powerful at surviving than any other substance.

Energy consumption per capita will likely fly off the charts during the singularity. This as well as all other natural needs, like water and clean air will be demanded. We will fight climate change, not by altering our impact on nature but with air conditioners and atmospheric scrubbers. These solutions require huge amounts of energy.

Computers alone have become huge energy hogs as they mine cryptocurrencies. Crypto has a virtual value by speculation only but is actually an unproductive energy use supporting an expensive online gambling game.

With the coming of AI, simulating an organic brain will require a hell of a lot more energy than any amount an actual organic brain consumes. Energy is both the price of progress and its measure and it's increasing exponentially.

We went from rolling rocks uphill to throwing rocks into outer space. Energy is a key part of the coming singularity, as technology will demand ever-increasing amounts. Power generation must now follow Moore's Law and double its production every few years.

In order for the human brain to keep up with the accelerating technology, it too must double its cognitive power every few years. With increasing technology, reality for the average human becomes equally complex and intricate. Humans interpret reality through a unique cognitive process directly connected to brain function.

The brain persistently simulates reality to predict upcoming events. When expectations do not align with actual perceptions, emotional responses are activated. Emotion drives increasing awareness, supporting rapid decision-making driven by intuition. This mechanism promotes probabilistic outcomes. It encourages attempts to predict and influence random events, which are often viewed as threats to survival. The natural inclination to fear an uncertain future stems from the brain's limited ability to simulate reality.

Random events often draw our disproportionate attention to an innate desire to know the future. Different cultures develop distinct perspectives on probability and the unfolding of future events. Some believe that random occurrences are governed by a higher, sentient being. This leads to the creation of various cultural beliefs contradicting reality such as the existence of luck. Others engage with probability as a form of entertainment. A fascination that has influenced the development of entire cities built around this weird human quirk.

Quantum mechanics fundamentally relies on probabilistic principles. It says that physical existence comes from an imaginary probability amplitude in a complex conjugant wave equation. We think of our universe as an expanding probability of leaving behind a four-dimensional space-time field. This expansion arises from the random fluctuations of a purely probable vacuum energy.

Energy began with the Big Bang. This event was an amazing surge of energy that stays permanently separate from the vacuum. It immediately breaks multi-dimensional symmetry with enough energy to sustain ongoing expansion. This expansion creates in its wake three spatial dimensions and one temporal dimension.

Expansion is also speeding up over time. This shows that space and time are free-falling into an empty energy well. It's as if the universe is being pulled into the ever-expanding wake of the Big Bang gravitational wave. We call it dark energy. When space untwisted into a Pythagorean geometry, it leaves behind warped areas we call dark matter. Space is not linear and whenever it becomes non-Pythagorean or warped, it contains positive energy and hence generates gravity.

Another intriguing connection between quantum mechanics and human behavior are physical measurements. In quantum mechanics, fundamental particles are unique. They have specific qualities, but we cannot tell them apart or predict their behavior individually. Nonetheless, they become accurately predictable when observed collectively at larger scales.

Similarly, humans are unpredictable individually. But within large groups or cultures, they tend to organize and become less distinct and thus become predictable in many respects. Quantum mechanics offers insights into how this phenomenon operates.

Quantum mechanics explains how fundamental particles and waves constitute physical reality. Similarly, civilizations emerge from the actions of individual humans, each playing distinct roles. Fundamental particles adhere to specific rules and quantized energy and momentum states. Humans, unlike particles, have many unique emotional states. These states are quantized and represent the simulation of our inner reality.

Emotional states are set by the culture that shapes early brain programming, conforming identity. Ultimately, reality depends on how we utilize information to feed our internal simulation of what we perceive as reality.

Another aspect of the coming singularity will be the power of our computing technology. The density of solid-state silicon circuits is rapidly approaching the scale of the neural clusters in the human brain. Neurons are electrochemically active cells. They join together to create complex networks in the brain.

A typical human brain has around one hundred billion neurons. These neurons form about a billion specialized units known as neural clusters or nuclei. I consider these neural nuclei as being a fundamental processing units or PUs. These clusters are the brain's basic computing units. Our current simulations of neural systems boast of having around a billion artificial neural networks. Hence, if our brain is aware due to its complex organization and programming, so would be an equivalent simulation.

Due to the complexity and extensive variations in PUs, I will only model them as generic network processors. Each one takes a specific set of inputs and produces a planned output or transfer function under some system control. This output can be processed further by other PUs functioning under a switched network, in our case, the neocortex. This is the detailed functional level I will use when comparing brain models.

A silicon-based digital Central Processing Unit, or CPU, acts like an organic neural nucleus. When it has firmware and connects to input and output devices under program control, it can perform similar smart functions. The expected singularity means silicon processors may shrink to sizes smaller than organic ones. This change could allow artificial neural networks to match the size and density of biological brains. It will allow for complete artificial replication of an entire human brain within the same or even a smaller space.

This is still a big challenge. Solid-state materials create heat that is thousands of times higher than what organic tissues produce. The human brain uses a lot of energy. It manages this well to avoid overheating and nutrient shortages. Sometimes, it even shuts down certain functions to save energy.

About a hundred years ago, quantum mechanics created a silicon-based technology. This technology is now essential to our daily lives. Manufacturers use silicon-based semiconductors in smartphones, computers, televisions, and automobiles. It didn't start a silicon era like the Stone, Bronze, or Iron Ages. But it changed how we share knowledge and information in a big way.

Silicon accelerates the sharing of information. Therefore, we live in an information age. Essentially, information generates knowledge, which is the primary force behind human progress. The pace of this is increasing at an exponential rate.

Humans speed up progress with quantum-like jumps in technological innovations. This is different from biological evolution, which occurs very slowly over an extended period.

As an example, iron was seen as rare and magical. People knew it only from meteorites and used it to make extremely rare weapons by heating and hammering. Its importance suddenly increased as people figured out how to make artificial meteorites through mining, refining and casting. This made iron more plentiful and cheaper, changing everything in a relative instant.

Technological advancements have defined human progress. For instance, our brains have complex neural networks. These networks lack pre-wired specific traits at birth, except for basic bodily functions. When children lack nurturing, they don't stagnate at some primitive version. Instead, they act like computers, totally useless without software running. We gain all essential human knowledge from outside sources. Cultural and social networks, information technology, shape our growth and skills.

Human brains require external programming before they can function effectively. After years of growth and learning, a human infant becomes a functioning member of society. They join the larger cultural network, which acts like a pseudo-neural thinking system. We live closely together, sharing information within a larger mesh network reminiscent of our organic brains. This virtual network, composed of human processing nodes, functions like a culture-computing system. Culture defines its symbols. It also defines the emotions we use to process information. This helps us sort information and make decisions intuitively and quickly. That's the main job of our brains; quick effective reaction to threats.

Human thinking depends on working together in shared cultures. This teamwork helps us survive and shapes our thinking. This gives us a big edge over our small-group ancestors. They were more rigid and less adaptable. Does relying on collective culture hurt our individual identity and free will? Does it make us just small parts of a bigger group that decides our survival? If so, it is an inherent aspect of our existence, and we must accept it. We are more akin to ants and bees than we might imagine.

The human brain depends on accurate information to maintain valid future simulations. New information can be easily shared among connected brains. Not all information is true or helpful. Tested and true information becomes knowledge. Extra work needs to be done to determine if information becomes knowledge. This processing generates emotional responses tied to what is learned. Truth is not needed for the brain to work, but it is crucial for producing survival, and continual progress.

Emotions help us organize knowledge for easy access. They play a big role in our awareness and identity. Changing emotions, whether by choice or outside influence, will affect basic brain decision-making. All human knowledge must be based on truth. This ensures it can be used effectively in technology. When true knowledge is applied successfully, it creates positive feedback feelings. Success leads to new knowledge. This knowledge is shared and processed within our connected culture. Emotions are not just human reactions or oddities. They are the key thinking tool for consciousness. They come before awareness and guide how we make decisions.

Humans possess a unique genetic mutation in their brains, setting us apart from all other hominins. This mutation results in a unique neuron structure in the neocortex. It has neurons with long, complex dendrites and axons that resonate between multiple states at their tips.

These special neurons link different brain regions. They create dynamic patterns that resonate throughout the entire brain. Humans have a greater density of these neurons compared to other hominins. This gives us more expressive language, traditions, rituals, music, and dance. As a result, we use and control a much wider range of emotions.

This control network acts as the brain's management center, enabling dynamic switching and sensing. It stimulates PUs to fire in structural patterns. This forms new circuit structures that process symbols through emotional connections. We can change how we think, but we need to relearn our emotional feedback systems if we are going to get fresh answers.

Emotions are feelings tied to our awareness. They help us process information and shape how we see reality. This feedback guides our future decisions. They signal when intuition suffices, reducing the need for expensive thinking.

Modern humans initially organize themselves into cohesive, self-programming cultures. Unlike early social groups dominated by a single strong male and a few females, contemporary groups are based on consensus among many adults. Males usually form monogamous pairs with females who together share social power. This helps to maintain social harmony and avoids violent fights over dominance.

Culturally connected breeding groups are now free from traditional reproductive and social limitations. They confer through symbolic language how to enhance their ability to plan hunts, teach tool use, craft woven items, and preserve food. This accumulated cultural knowledge continually expands their skills and survival strategies. Enforced communal living changes natural conflicts about reproduction into mutual empathy. This connection helps bind the group together and fosters their enhanced survival.

But this strategy has vulnerabilities. Unscrupulous individuals can manipulate group planning, turning it into parasitism. Humans often deceive and make promises to manipulate others. This is different from animals, which do not use deception for general survival.

People do this mainly to control the culture for their selfish wants. This is akin to a networked computer being hacked by pirates demanding hostage payment. The human brain doesn't have defenses like firewalls or virus protection. This is mainly due to our strong emotional ties to our group, bypassing our natural fears that normally drive individual decisions.

On one side, we have rigid cultural groups striving to control the future by shaping environments to meet their needs. On the other, we have humans who consciously need to adapt. This conflict helps us change our cultural thinking when faced with new challenges. This is accepted when such change depends on maintaining a sense of security and belonging within the group.

Understanding a machine, much like the human brain, requires examining its fundamental workings. This means knowing the basic physical principles and natural laws that influence its design and operation. Without this basic knowledge, you cannot understand or change how it works.

When the mind understands a phenomenon or mechanism, it builds a symbolic neural network to deal with it. This network simulates the process. The visual network turns sensory observations into a symbolic virtual imagery. This lets us manipulate and project actions into the future. This ongoing simulation of personal reality consists entirely of symbols. It turns them into clear visuals and time-based experiences through awareness. New ideas emerge from the interaction of different simulations of the same reality, rather than from nowhere.

When people try to stop progress, thinking we should limit technology, they depend on willpower that goes against nature. Progress is driven by the need for survival, a natural and continuous process. It's a mistake to believe that survival depends on force, either physical or mental. History shows that these strategies ultimately harm humanity and other valid cultures. Relying only on force usually leads to a general downfall.

Instead, we should focus on knowledge and technology to ensure our future. When faced with an insurmountable force, some humans will adapt by acquiring better fitness, while others may not. Those who adapt will gain an edge. This could lead to the rise of new cultures, races, and even species.

Tech advancements can boost certain genetic traits. These traits quickly get added to DNA in ways that go beyond just reproduction. Horizontal Gene Transfer (HGT) lets cells swap DNA in real time. This means genetic changes can happen quickly. For example, traits like new brain wiring or changes in chemical neuro-transmitters can be passed on in just one or two generations. We can potentially engineer improved brains swiftly if necessary.

A pinch-off event happens when a species splits into two unequal branches. This must occur from a common ancestor. One branch will always have a survival advantage because of competing survival strategies. Human technology has greatly improved our survival skills. This success comes from big leaps in knowledge. We can share this knowledge, keep it for future generations, and use it to make life better. It helps us live longer, enjoy more free time, and improve our health. This progress compels us to learn more, do more, and expand our physical and mental environments—from oceans to outer space. Humans uniquely utilize information to reprogram expectations, making us the accumulating information animals. Our interconnected brains and virtual networks require this continuous influx of knowledge in order to survive.

The human brain is the most advanced computing machine known today. It can be programmed to simulate any physical process perceivable by the senses. You can train it through knowledge to explore *'what if'* scenarios based on natural laws or logic. It models natural actions over time. The more it learns from other brains, the better it predicts future outcomes. It acts like a playground for creativity. It simulates symbolic worlds and projects them into the future under intuitive rules. Then, it helps choose emotional reactions to get the results we think we need.

Humans can predict the path of a curving projectile through instinctive understanding. They do this by observing, simulating, and coordinating. There's no need to solve complex differential equations. They just catch the ball.

They can strengthen their bodies and improve agility through practice. By repeating actions, they train their neurons to fire in specific ways when triggered by their senses. Great musicians and artists can reprogram their brains at will. This skill helps them create lasting changes in how they connect and respond emotionally. Desirable emotions act as positive feedback for successful learning.

The impending singularity signifies a pivotal transition for humanity. It promises to free us from needing to work for a living. Robotics and AI will handle production and service jobs in every field. Freed from the burdens of physical survival, humans will be able to pursue personal interests and passions. Once the obligation to toil for basic needs is removed, we will be truly free to explore and enjoy what brings us personal fulfillment. We will become the super self.

The fight for economic control will fade. It will be replaced by a system that rewards those who best meet our shared needs for fulfillment. With advancements in technology and AI, life becomes increasingly self-directed and self-enriching. We will have the power to create and shape our own realities and destinies. We will break free from cultural limits. This lets us program our minds and build unique virtual worlds for just our private amusement.

Applying AI to the complex chemistry of life will enable us to master our physical health, maintaining vitality and longevity. Smaller robotic and nanotech devices will reach cellular and molecular sizes. This will enable medicine to deliver precise, personalized treatments. Maintaining control over AI agents is essential to harness these revolutionary benefits responsibly.

It is hoped that this book will help the reader understand these issues better. It explores life's evolution and how it led to our current situation. It will examine how the human body and brain function at a logical, machine-like level and why. We'll explore how our interest in chance and probability affects our adaptability to an uncertain future. We will also look at social institutions, beliefs, and technologies in general.

A new kind of human needs to appear. They must be more aware and better adapted. This will help achieve the long-held goal of becoming divine through wisdom. Personal growth through emotional mastery and self-programming will help people survive the singularity. This can lead to powerful changes in both body and mind. In the end, we might reach a state similar to the Greek gods.

We could live life as a virtual reality show, fueled by our selfish wants and victorious struggles, with the universe as our stage. With time, knowledge and wisdom grow. Our virtual realities might bring us to personal nirvana. This can be a lasting state of complete awareness, humbling empathy and supreme unity with the universe.

Chapter 2: How it all started.

When I was a teenager working on my family's small self-subsistence farm in Oregon, I enjoyed listening to the radio. Connecting to a larger, more exciting world while shoveling cow manure and polishing eggs helped me stay focused. It made the long, boring hours of hard work pass quickly. But my family was poor, so I couldn't afford to buy a radio. But whenever I took our garbage to the dump, I'd look around and maybe pick up an old radio or two that had been tossed. I'd take them home and, in the evening, tear them apart to try to learn how they worked and maybe learn how to fix one.

After many failures, I discovered one with a simple broken wire. I easily fixed it with soldering, a skill my father taught me. I keep this up by finding and fixing television sets and just about any other electrical appliance I needed. It also helped me find a career. Later, I fixed network issues in MCI's new digital communication system. What culture forces me to learn as a youngster gives me the skills to rise above that culture and into an entirely new one.

To fix anything, first you have to know how it's supposed to work and the basics governing how it works. Then you isolate the problem by applying a process of elimination and some simple logical analysis. In theory, anything that once worked can be made to work again. That's the key to technology. It's totally repeatable and so, so logical. But if you understand it, then you have a chance at making it work the way you want it to. If you don't understand it, then you'll probably be calling *"the man"* to come and fix your technical stuff. You will be at a great disadvantage compared to those who understand technology, who can fix things and make them work.

If I am going to provide the basic knowledge and tools to take on the singularity, first we'll need to understand how this whole mess works. If you don't know the basics, how do you know when it's broken? To tackle this issue of pure cosmic awareness, we need to grasp everything. We must understand how things work, why they work, where they came from, and where they're headed.

Along these lines, we'll look at organic life in the context of the cosmic universe driving it. We'll understand where life comes from, how it works, and where it's going. We'll explore how life began and changed the Earth. This journey will lead us to human evolution, awareness, technology, and cultures. All of these are necessary for dealing with an information singularity.

Next, we'll explore intelligent machines, or computing machines. We'll focus on those with smart networks. We'll learn about their origins, their purpose, and their future. Finally, we'll consider how organic and artificial intelligence might converge. This could lead to a major extinction event, a possible pinch-off event for the present-day human species.

It's an issue big enough to require huge perspectives. We need to step back and view the whole picture. This includes everything from the Big Bang to the Cosmic Microwave Background. It spans the basics of math, physics, and chemistry. It also covers geology, biology, zoology, anthropology, and archaeology. We must figure out how we ended up living in this strange world. We dominate all other life but feel disconnected from it. It's hard to understand why it exists when we're not watching.

We need to examine quantum mechanics closely. Its core principles show how this natural phenomenon is a fundamental description of reality. It works at all scales. It explains both large events and tiny details.

We'll look at what intelligence means. This includes thinking, reasoning, and making decisions. We'll see how it's accomplished with a complex and versatile neural network that has evolved naturally. Next, we will look at how digital computers can mimic neural networks, creating an artificial brain. We'll compare artificial intelligence and organic intelligence. They might just be different forms of the same phenomenon.

Don't worry, functionally understanding doesn't require a lot of math or formal science. It does require an imagination able to represent nature with the use of familiar symbology and incomplete imagery. Unusual perspectives of reality will be needed to understand nature and how it functions at all scales.

For instance, when I think of atoms arrayed in a solid, I think of a matrix of balls with sticks connecting them in some regular 3-dimensional array. Better, I can imagine the sticks holding them together as springs that can vibrate in harmonic modes mimicking the way atoms move in a solid. Now I can imagine sound waves moving through it and can see how this kind of sound is different than the ones in air. It is far more complicated than air waves in that all three directions are coupled giving it properties beyond simple pressure waves.

I can even imagine an irregular atomic array where the springs align in tubes or sheets, mashed together in irregular patterns. They might seem random at small scales, but predictable patterns emerge at larger scales. The balls are atoms and the sticks are where electron wave functions are shared between atoms.

The point of this is to train your brain's neural network reality-simulating machine to expand its normally narrow understanding of reality. We must broaden our animal scales to include other realities at other scales. Picture a glowing, soft substance moving in odd, 3D football-shaped flower patterns. It surrounds a tiny point of attractive force; the nucleus. The electrons do not just fall into and glue themselves to the tiny proton, as intuition might predict. It simply cannot exist in a state like that. Nature forbids the collapse of the universe.

The lowest state and the closest an electron can get to a proton is in a spherical shell surrounding it at about an Angstrom or 10^{-10} meters. A proton is about 10^{-15} meters.

But the electron in an atom becomes a negatively charged smear or cloud of probability at an average distance of 100,000 diameters of the proton. Imagine a near point charge of positive electric force attracting a negative charge. As the

A quick review of numbers and how we represent them.

All numbers that are real, can be represented by:

0,1,2,3,4,5,6,7,8,9 are number symbols so we can count. But when we get to big numbers like:

one billion, which looks like this, 1,000,000,000

I'm lazy and get tired of writing zeros and keeping track of commas, or if you're european, periods

So, instead we write a billion as:

1.0×10^9 which says the number is one times ten to the ninth, or nine zeroes.

Our universe is very large and our building blocks are very small, so we need to represent our numbers this way which makes easy writing and computing.

Ex. $2.0 \times 10^{-3} \times 2.0 \times 10^7 = 4.0 \times 10^4$

Rem: exponents add when multiplying numbers

electron gets close to the proton, it must give up energy in the form of photons until it has no more energy to give up.

Hence, the electron settles into a stable ground state that gets no closer to the proton or lower in energy. This becomes an atom of hydrogen, the most common and simplest element in the universe. And this isn't even close to the smallest end of nature's scale.

Reviewing metric vocabulary: kilo means a thousand and milli means one one-thousandth. ($1000 = 10^{+3}$ = kilo, $1/1000 = 10^{-3}$ = milli) Going up, mega is a million, or 10^{+6}, giga is a billion, or 10^{+9}, tera is a trillion, or 10^{+12}, and so forth. While going down, micro is one one-millionth, or 10^{-6}, nano is one one-billionth, or 10^{-9}, pico is one one-trillionth, or 10^{-12}, and so on. If this is unclear, search for explanations on Wikipedia or the internet for deeper understanding.

Today, after just a few centuries of serious science, we can see a universe that is about $1 \times 10^{+27}$ meters wide. On the other hand, the smallest piece of matter we can single out from nature is the quark, which is about 1×10^{-30} meters in size. We live in a cosmos that we can directly observe over a scale of about 10^{+60}. This stretches from the strange vacuum lurking at the Planck scale to the vast universe filled with all matter, energy, space, and time.

I can't fully comprehend such a large number. However, when I step back, the big things catch my eye, and the small details blend into the background. Readers need to make the

Common Unit	Size (10^n m)		
	-35		Plank Length $1pL$
ym	-24	v_e	Neutrino
	-23		
	-22		
zm	-21		
	-20		
	-19		
am	-18	e^-	Electron
	-17		
	-16		
fm	-15	p	Proton
	-14		
pm	-13		
	-12		
	-11		
Å nm	-10		H2O
	-9		
	-8		
μm	-7		Proteins
	-6		
	-5		
mm	-4		Cells
	-3		
	-2		
m	-1		
	0		Human
	1		
km 1	2		ISS
3	3		
5	4		
7	5		
AU	6		Earth
1	7		
2	8		Sun
3	9		
4	10		
Y	11		
1	12		
	13		
	14		
	15		Solar System
4	16		
	17		
6	18		
	19		
8	20		Milky Way
	21		
	22		
10	23		
D_0	24		
	25		Observable Universe
	26		
	27		

attempt to appreciate the big view. This will help understand the overall picture of reality. It shows how natural forces change things over time and across different spaces.

For instance, there are about 10^{+23} atoms in the space of about a cubic meter. This is a huge number to think about. But when you average many objects that are close together, the differences in anything measurable shrinks. This makes it seem like you are dealing with just one object instead of countless individual ones. Our instruments are large. They need many atoms to work, so they only measure averages from big groups. At the atomic level, atoms move randomly and chaotically. But when we look at a larger scale, their behavior becomes predictable through averaged probabilities.

Large number probabilities play a key role in how we perceive the reality we live in. Humans have a strong awareness of the future and its constant, unforeseen influence on our conscious reality. Nature's unpredictability fascinates us. We invest a lot of time and resources just thinking about it. We play with it for fun and entertainment.

We try to predict the future and feel rewarded when we guess correctly. The only thing is, sometimes it's purely random and, by definition, unpredictable. There is no way to consistently win. And yet, games of probability continue to fascinate us because we think we are special and subject to something called personal luck.

Humans especially take note of seemingly rare events that seem to come and go magically without rhyme or reason. When something extremely rare happens, they feel unique. It's like the universe or God has set them apart. Fortunate or unfortunate, happenstance turns into, *"God did it!"* Our culture shapes us to think this way. We are tied emotionally and semi-rationally by the *"cultural conformity addiction."*

Big number probability tells us that no matter how improbable an event, given enough opportunities, a.k.a. phase space for it to happen in, only makes it a sure thing. With so many aware people pushing their limits every day, anything can happen. For the random person watching events unfold, it's either a miracle or a disaster. It all depends on how they see things. From a universal reality standpoint, such rare events are just normal and should be expected. Remember, the odds of winning the lottery are roughly the same as getting run over by a bus, so choose your luck carefully.

In the beginning, there is no such thing as a beginning or an end because time hadn't been released or expressed yet. All the universe's energy was in a singular state of a pure substance that does only one thing. It expresses probability according to the laws of quantum physics. If the chance of an energy fluctuation or big bang appearing from nothing reaches one, it happens. There's no time yet to measure when it doesn't happen. So space-time bursts from a singularity. It unfolds, untwists, and unwinds into a continuous three-dimensional vector space coupled to a temporal dimension, forming a Pythagorean field or a Lorentzian space.

String supersymmetry starts with 10 dimensions. Then, it spontaneously breaks a symmetry. This creates a 3-dimensional space field linked to linear momentum. It also involves an imaginary time dimension tied to energy. Energy and time are entangled by Heisenberg's constant; h. Space and time conservation means that tightly packed quanta of folded space must release its curvature as intrinsic spin and it's linear space as linear momentum.

The intrinsic spin field decays into the Higgs field, gaining mass and quantized angular momentum. This then leads to the creation of fundamental particles, following the rules of quantum mechanics. These rules are best represented mathematically by using Hilbert spaces. Linear space is continuous, like the number line. Hilbert space, however, focuses on discrete numbers. These are like the sides of a die or vectors that point to different locations in a spherical space. Quantum mechanics says one key thing: nature exists in small packets and waves. These waves resemble vectors in a multi-dimensional energy space.

Space at the beginning is warped, similar to a tightly knotted string. It exists as an energetic singularity. It might also exist as a combination of all its lower energy states. To achieve this, tightly folded space or supergravity needs to unfold into two types of space.

One is linear space, shown by Newtonian force vector equations. The other is spin space, linked to rotational Hilbert spaces. This relationship creates one of the universe's great constants, π (pi), which connects linear space to curved space. Newton's linear equations are only valid with linear space. But real space consists of both linear and non-linear parts.

Einstein modifies Newtonian mechanics with relativity. If something warps space-time, it disturbs the balance between time and its three spatial dimensions. This creates gravity. Accelerating matter warps space, also creating gravity. Regions of natural space-warps left over from the Big Bang attract mass, making them appear as dark matter. Attract enough matter, and when it warps space completely around itself, effectively cutting itself off from linear space, a black hole forms.

Although we're talking about the vast universe, these odd traits come from the basic rules of nature in quantum mechanics. This makes quantum mechanics the math of everything, not just the tiny particles where it was first found and dominates.

The universe starts as an entangled 4D space-time expanding with a positive force. Meanwhile, everything left behind unfolds into a linear 3D field, plus the 4th dimension: time. Gravity pulls in matter and energy from this field until it becomes strong enough to overcome the quark forces within the proton. They mix with other bare quarks, fusing some together into lower energy states forming the first 26 atomic nuclei. A star is born burning quarks and releasing the excess energy as photons and more quarks. Remember, mass appearing as matter is just another form of energy and vice versa.

Let's do a quick physics review about force and energy, in case you missed that part in high school. Three natural forces exist: the first and foremost attracts all matter to other matter. Gravitational attraction is the longest-ranging and weakest. It controls the shape of the universe.

Other forces are shorter in range. However, they become much stronger than gravity as distance decreases. After gravity, the next force is the electromagnetic force, or EMF. It has two types of sources: positive and negative Coulomb charges. The most famous Coulomb charge is the electron with a negative charge. This creates two forces: attraction between opposite charges and repulsion between like charges. Electronics is based on moving electron charges around circuits made of conducting materials.

The weak force is a special form of the EMF. Certain massive and unstable particles related to photons carry it. This force controls slow nuclear particle decays. These decays are crucial for stars by creating all the chemical elements we know. It's hard to understand, and Feynman won the Nobel Prize for calculating it. He showed that at high energies, the EMF and weak force are the same force, called the electroweak force. Experiments confirmed the discovery of the massive Zo and W+/- particles carrying this force at CERN in 1983 and later at Fermi Lab in Illinois.

Finally, the gluon force, or strong nuclear force, which holds the quarks together in the proton, cannot exist as a bare direct field like the others. It only exists in nature as a second-order fringe field that leaks out from between the bound quarks inside hadrons. We need eight numbers to describe the gluons' energy field. This follows the rules of quantum mechanics and a specific math called Lie algebras. Trying to pull two bound quarks apart stretches the gluon force. They don't just break into two particles. Instead, the energy needed is so great that it tears a hole in space. This lets vacuum energy unleash showers of exotic particles from what was once a stable particle.

The laws of physics play out in the early universe until stars begin lighting up the dark for the first time. Energy is a clear part of nature. It cannot be created or destroyed. Instead, it changes forms and is always conserved. It is a fundamental law of nature, similar to the conservation of momentum, mv, and the conservation of electric charge, q. When things are conserved, counting becomes vital. It ensures that balance is kept across large distances. That's why math works so well for predicting nature. Nature uses number logic to count and create balance. Because of this, math often mirrors nature in a precise manner.

Let me backtrack a bit here again and discuss how the electric fields that hold atoms together can form waves we perceive as heat and light. Electrons differ from quarks. They are lighter and have no intrinsic size. They flow easily through materials with loosely bound electron clouds, like metals. This cloud creates a quantum fluid that lets other electrons move through with little disturbance. Electrons are conserved and can only move if an electric or magnetic field creates a force, EMF, and they are free to move in response.

An electric field exists whenever a positively charged atomic nucleus is separated from a negatively charged electron. Together, their attractive force makes up all matter in the universe. It's why we can't push our hands through a table even though both of these objects are 99% empty space. We are pushing our electrons against the table's electrons, and they just say *"no freaking way!"*

Like charges repel each other, while opposite charges attract. However, quantum mechanics limits the stable stationary states that can exist between them. Nature has strict rules about two field sources getting too close to one another. Think about it. This would lead to all matter collapsing in on itself. It would release enormous energy, resulting in a massive explosion similar to the big bang. Quantum mechanics says there is a minimum energy state that electrons can exist in and no others lower, resulting in stable matter.

When an electron moves with its electric field, it creates a magnetic field. This field wraps around the path of the electron's movement. When electrons move back and forth, like sloshing water in a pan, they lose energy. A force retards their movement turning some of its energy into photons. These photons match the frequency of the oscillating electrons. It's as if the electrons are moving through a liquid that pulls energy away in the form of a wave. Electric and magnetic fields combine at right angles to form a single energy field, a moving electromagnetic wave.

It's an electrical energy pendulum. Energy first stores as an electric field. Then, as it collapses, a magnetic field builds, soaking up the energy. This field then reverses, sending the energy back to an increasing electric field. It keeps cycling like this until it interacts with something else.

Radio waves are created by sending large electric currents through a conducting antenna. The electrons move up and down at a resonant frequency. This frequency matches the antenna length with the electromagnetic or radio wavelength. This electromagnetic wave radiates away, carrying a unit of photon energy.

Dynamically, an electric field is formed when electrons are pushed to excess at one end of an antenna or the other. A magnetic field is produced from the moving current flowing between the two ends of the antenna. These two waves oscillate at the same frequency but are 90° out of phase and at right angles to each other. As the magnetic field decreases in amplitude, the electric field rises, and so forth.

They move in a back-and-forth, see-saw grip, always along one direction of travel. This is a photon particle. It carries energy and angular momentum based on its wavelength and travels at the fastest speed allowed by nature. It goes for a timeless ride through space, rotating its phase clock until it hits something charged, like another electron.

An electron reacts to the colliding wave by absorbing its energy. This can make it hotter or speed it up. It may also emit another photon in a different direction and different frequency.

The speed of light is not so much a speed limit but a property of linear space that simply can't allow physical changes any faster. The speed limit for sending information through space exists, but there is no limit how much information can be passed. It only says that the higher the energy the more information it contains.

Lowest energy photons are radio waves. Higher energy photons become microwaves and eventually infrared heat. Higher still are visible light, followed by ultraviolet, x-rays, and finally gamma rays. As energy per photon increases, the wavelengths get smaller and more localized. They start to behave like solid particles.

This allows them to penetrate other fields and cause significant damage to any particles in their path. Photons are both particles and waves. Mostly waves at low energies and frequencies while mostly particles at high energies.

Actual wavelengths can range across the entire distance scales of nature as illustrated above.

$E = hf$ and $f = c/\lambda$ are key photon relationships. Here, E is the photon energy, h is Planck's constant, f is frequency, c is the speed of light, and λ (lambda) is the wavelength. All electromagnetic waves obey these two equations. Also, all waves must fit the general wave equation: $f = v/\lambda$. Here, v is the wave's velocity, and λ is its wavelength. It can be the speed of sound for air pressure waves or the speed of a tsunami wave in the open ocean.

Plank's constant is crucial. It shows how much energy is in a photon at a certain frequency. It also defines the smallest distance where space and time connect, known as quantized space. It has a value of 6.6×10^{-34} Joule – sec. It is shown in the distance scale illustration on the extreme top. In other words, there is a smallest piece of space that can exist in reality and nothing smaller. It is the fundamental unit of the quantum world, giving us the size where all natural variables get fuzzy and become unknowable.

If I use a term or word, you are not familiar with and I haven't defined, then look it up on the internet. Every day, people invent a lot of technical words, so we have to keep up by not letting bullies push us around with words we don't understand. We live in a unique time. Now, everyone has access to all information. This is true, no matter their thinking skills.

We're free to explore any topic of human knowledge. When we do, we often find that many others have spent years studying the same thing we want to learn about. Be strong and learn from those who came before. These people dedicated their lives to discovering and sharing new knowledge. Every generation understands and sees more from the advantage of standing on the shoulders of giants.

It's a key ingredient that we, as intelligent humans, must have. Any and all knowledge, useful or not, should be common knowledge, like a free lending library. It once was, we lost it, and it's finally returning. Humans now routinely carry around the world's sum of all current knowledge in their pockets.

However, we must still be the last responsible person in the network to determine if the information posing as knowledge is valid and useful, or false and misleading. You can only trust those who clearly know best, but even then, check your sources. It's a duty of every aware brain to check on truthfulness. Verify your facts, or you might face the consequences of making risky life-preserving predictions based on bad information.

Remember, for any thinking machine, bullshit in equals bullshit out. The brain doesn't really care until it has to eat some of it and then it realizes there is a difference between shit and Shinola. If we do not do our due diligence and make sure the information we are seeking is true, honest, and verifiable, we risk losing everything including life itself.

You can't build a brain that predicts reality successfully with only falsehoods, lies or simply selfish wishes. It needs more than good feelings to handle life's random challenges. Nature just doesn't work that way, as much as we might like to think it should. Nature is not fair, but it is flexible.

Use this uniquely human attribute: if you don't know, find out. When you have a question, the first step is to look it up online. Get the latest information, both for and against. Verify its truthiness first. Then, add it to your collection of truthful knowledge. This will enhance your tool belt of correct information, actual reality, controlling power, and creating new things. Nothing real comes from bad knowledge.

Stars light up from nuclear energy when their cores reach a high enough density and it gets hot from all the energetic stuff falling into it. Young, virgin stars quickly fuse quarks into heavier nuclei. This process releases huge amounts of energy and forms the lighter elements, all the way up to iron. Any heavier elements beyond iron require adding energy instead of releasing energy. Building elements from a hot quark soup can unbalance plasma thermodynamic forces and gravity. This sometimes leads to a massive explosion called a supernova. In just seconds, this explosion releases enough energy to create significant amounts of elements a lot heavier than iron.

The exploding Nova collapses into a neutron or cold quark dwarf, continuing to pump out energy as particle beams

PERIODIC CHART

- Alkali Metal
- Alkaline earth metal
- Lanthanide
- Actinide
- Transition Metal
- Other Metals
- Metalloids
- Non Metals
- Halogen
- Noble Gas

1	2	3	4	5	6	7	8	9	10	11	12	13	14	15	16	17	18
H																	He
Li	Be											B	C	N	O	F	Ne
Na	Mg											Al	Si	P	S	Cl	Ar
K	Ca	Sc	Ti	V	Cr	Mn	Fe	Co	Ni	Cu	Zn	Ga	Ge	As	Se	Br	Kr
Rb	Sr	Y	Zr	Nb	Mo	Tc	Ru	Rh	Pd	Ag	Cd	In	Sn	Sb	Te	I	Xe
Cs	Ba		Hf	Ta	W	Re	Os	Ir	Pt	Au	Hg	Tl	Pb	Bi	Po	At	Rn
Fr	Ra		Rf	Db	Sg	Bh	Hs	Mt	Ds	Rg	Cn	Uut	Fl	Uup	Lv	Uus	Uuo

La	Ce	Pr	Nd	Pm	Sm	Eu	Gd	Tb	Dy	Ho	Er	Tm	Yb	Lu
Ac	Th	Pa	U	Np	Pu	Am	Cm	Bk	Cf	Es	Fm	Md	No	Lr

and radio waves. Dust and cold material from explosions are drawn to other gravitational centers. Over time, they combine with more hydrogen to form second-generation stars. These stars are heavier and denser, so they burn much slower. They are also rich in elements, as is all the leftover supernova material showing up as planets and asteroids. There's a lot of space debris caught in our second-generation sun's gravitational field, with the Earth just one very small piece of it all.

Atoms with increasing numbers of protons make up the elements from which all chemistry is derived. The periodic table of elements is one of the most powerful tools for predicting chemical reactions. In its simple structure lie the secrets to building a living universe from just simple elemental pieces.

Some early giant stars do not go nova and instead collapse into a different physical state, entirely quantum in nature, called black holes. Black holes are formed when the leftover star core of bound quarks is so dense that its gravity squashes it down to below the Schwarzschild radius. This is the radius as a function of mass where the escape velocity exceeds the speed of light. What goes up must come down, except if you throw it fast enough, then it doesn't. Every massive object with gravity has a unique escape velocity for anything shot up and not returning to it eventually in a giant parabolic arc. Black holes can't cool down. They warp space tightly, trapping photons inside. So, hot photons can't escape.

These massive monsters have strong gravity and shred everything near them. They devour stars that stray too close. These heavy objects sweep in ordinary matter like stars, dust, and gas over long distances. They form clusters and galaxies that we can see through telescopes.

Galaxies come in all kinds of shapes and sizes, usually depending on the size and age of the black hole at their center. Larger ones usually form two spiral arms extending out tens of thousands of light-years in a symmetrical flat disk shape. This shape is characteristic of a chaotic process described mathematically by Mandelbrot sets.

A light year is the distance light travels in a year. It's a damn big number but not so big for our universe, which is about forty billion light years across. For a photon from the big bang, many of which still travel as cosmic microwave background radiation, time hasn't moved at all.

As space grows, its wavelength stretches. It shifts from hot ultraviolet radiation to cold microwaves. It now sits at about 160 GHz. This frequency is just above most commercial microwave systems. It also represents a temperature of 3 degrees above absolute zero, which is -273 degrees Celsius.

Galaxies are strange because they don't follow the same orbital rules as smaller systems. For example, our solar system, which includes the sun, planets, and moons, behaves differently. There is 70% more gravity buried throughout the galaxy than the amount we can see as glowing stars and dust. The stars orbit the center all at the same speed. It's like they're caught on a carousel. This is similar to how all the mass inside a planet moves at the same rotational speed. In normal planetary orbits, the orbit is much bigger than the gravity source. So, the velocity depends only on the square of the distance from that source like the planets going around our sun. Galaxies must have a lot more mass to cause this.

Many kinks or ancient space warps likely remain hidden in linear space after the big bang. This explains the extra non-glowing gravity found in deep galactic space. Dark matter might just be leftover spatial eddies from straitening Euclidean space. These eddies create gravity through their intrinsic curvature.

For our purposes, we tighten our focus at the point in the evolution of the universe where life arises on this planet, a heavy element planet full of supernova dust. Life seems out of place with stars and galaxies. It seems to lack a reason to exist in a universe filled with countless stars, black holes, galaxies, and vast clusters of galaxies.

All matter consists of around a hundred elements shown before in the periodic table. These elements share a common structure of a nucleus made of protons (p) and neutrons (n) and surrounded by clouds of electrically neutralizing electrons in tightly bound energy states.

Protons are made of two up-quarks (u) and one down-quark (d), while neutrons have one up-quark and two down-quarks. So, p = uud and n = udd. The quark charges work like this: the uud combination gives the proton one positive electric charge. In contrast, the udd combination leads to no net electric charge, which makes neutrons neutral.

Now, let's look at these quarks more closely. The up-quark has a charge of $+2/3$. The down-quark has a charge of $-1/3$. The two up-quarks in the proton add up to $+4/3$. The single down-quark subtracts -1/3. This leaves the proton with a total charge of +1. It is strange for a unique force field to connect a simple electron with integral charges to nucleons that have fractional charges. These are two totally different particles that have no reason for a common yet different electrical charge makeup.

Protons and neutrons, quarks, are bound together in a nucleus by the residual gluon force. If you think two charges are complicated, consider that gluons have eight. But back to electric charges.

The combined positive electric charge of a nucleus attracts the negative electrons. These electrons settle into specific energy levels around the nucleus. It's similar to standing waves on a drumhead or overtones in a hollow chamber. They resonate and form increasingly complicated shapes of probability around the central nucleus.

When a nucleus has six protons and six neutrons, it creates a very special element. This element bonds its electrons in a unique three-dimensional way. A carbon atom is unique. It can form complicated and extremely large molecules with nearly all other elements. This allows for a vast range of complexity and size required for life.

Carbon has two inner electrons that are tightly bound. This means no additional electrons can easily enter or leave its ground state. In the next higher up level in energy, eight electrons are allowed to share the same energy state, and carbon has exactly half that amount.

The four outer electrons in a carbon atom are loosely bound. Two of these can share with two other carbon atoms. This forms a continuous chain. The remaining two electrons can bond with almost any other element that wants to accept them. The bonding geometry offers countless possibilities, combinations, and variations. This leads to nearly endless configurations. Carbon can bond into large hydrocarbon

```
     H           H H         H  H  H   |    H  H  H  H  H  H
     |           | |         |  |  |   |    |  |  |  |  |  |
   H-C-H      H-C-C-H      H-C-C-C-H   |   -C-C-C-C-C-C-
     |          | |         |  |  |    |    |  |  |  |  |  |
     H          H H         H  H  H    |    H  H  H  H  H  H
                                       |
  Methane      Ethane       Propane    |     Hydro-Carbon Chain
```

chains with unique shapes and physical functions. This complexity creates many possibilities. It makes even the rarest event, life, more likely when chemical conditions are right for hydrocarbons.

This is an example of a special protein. It's a carbon chain molecule that can endlessly repeat itself. It does this using

simpler carbohydrate building blocks.

In this case, it's like a chain where the links repeated are the amino acids that make up these giant twisted protein

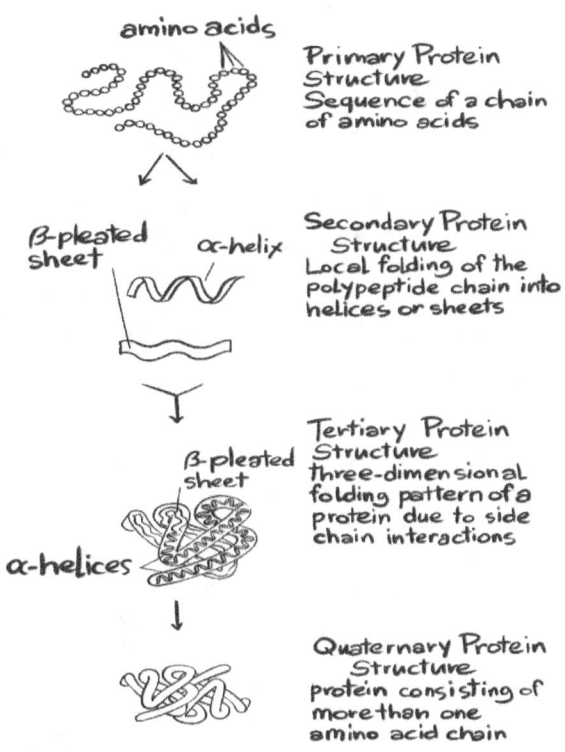

molecules, the seeds of life. The downside to this class of molecules is that they are very sensitive to energetic light and temperatures. Some bonds can break easily when hit by a fast-moving water or nitrogen molecule. This happens in high-temperature environments. Even low-energy photons like infrared can break chemical bonds and split fragile molecules. As a result, these molecules can't carry out their important life chemistry tasks without a stable and moderate environment.

When the sun and Earth reach about one and a half billion years old, the Earth cools down. This cooling happens as water vapor evaporates and condenses. The process releases a lot of infrared energy into space. As a result, the Earth cools even faster and stays cool. This is due to the unique structure of water. It is an oxygen atom that has two electron bonding spots, 90° apart. You might wonder why they aren't just on opposite sides and be 180° apart, but oxygen is different from carbon. Find it on the periodic table and you'll see how different it is.

Two hydrogen atoms, each with one electron, occupy the vacant spots on oxygen. But as they do this, they sense they are too close. This causes the hydrogens to repel each other a bit. As a result, the angle between them in the H_2O molecule is about 104° 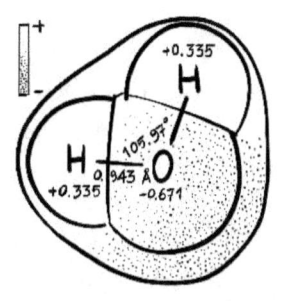 instead of the expected 90°. This makes it look like a tiny boomerang, creating a small electric dipole. A tiny electric field surrounds each water molecule like the magnetic lines of force around a bar magnet. This field gives water unique properties that set it apart from all other small molecules. It can affect the other major force holding the universe together, electromagnetic fields. This makes it one of the strongest chemical solvents in the universe.

This weird shape can have all kinds of quantized energy states of spinning, vibration, and bending. The electron bonds are squishy, flexible, and can vibrate like Jello. When they vibrate, they couple to the electromagnetic spectrum, absorbing or radiating energy. But only in the low-energy infrared spectrum. Water can absorb a lot of energy from low-energy heat photons. It does this through its vibrational and rotational modes. These modes have many quantized energy levels. Water then stores this energy and can reradiate it later. This process allows water vapor to behave like a greenhouse gas.

Water is incredibly useful as steam. It can store and release a large amount of energy when it changes from liquid to gas and back. With the fact that it also expands when going from a liquid to a solid makes this tiny molecule very special indeed.

But when the earth's average temperature drops to below 100° Celsius, the boiling point of water, rain happens and the oceans result. When water turns to vapor, it creates clouds. These clouds form tiny droplets. This happens because water molecules have a small electric field. They can line up like little bar magnets. They gather at the surface, forming a skin.

Inside, many more molecules are packed together. These tiny water droplets appear white. They reflect the sun's heat back into space, acting like a mirror. This helps cool the planet by providing shade and radiating thermal energy.

But the carbon equivalent of the dipole water molecule, like CH_3 (methane) or CO_2 (carbon dioxide), doesn't have a dipole. They stay clear in vapor form. These gases absorb and reradiate infrared photons. This traps heat in the atmosphere, warming the Earth's surface.

Earth gets extra heat from heavy radioactive elements lingering in the core disintegrate. These elements break apart because their large nuclei are packed with too many protons. The protons push against each other, eventually overcoming the gluon field, which makes them unstable or radioactive. This means there is a greater chance of the atom splitting into smaller atoms.

As an example, naturally occurring uranium used to make early nuclear reactors has a half-life of about four and a half billion years. This just happens to be about the age of our solar system. The Earth has lost half of its original uranium atoms as a source of some of this core heat. Every half-life period, half of the remaining atoms disintegrate into smaller atoms. This process releases energy held by the gluons into lighter nuclei, and heat. This heat helps keep our iron-rich core molten and spinning. A spinning iron core creates a huge dynamo generator of electric currents flowing in a circle. Current flowing in a circle around the core creates a magnetic field that extends from our core all the way out into deep outer space.

This large but weak magnetic field surrounds the Earth. It protects it from nearly all the streaming hot plasma particles, primarily electrons and bare protons, flowing out from the Sun. When an electrically charged particle's motion crosses a magnetic field line at an angle, it feels a new force. This new force is at right angles to the plane of the two directions.

It is deflected left or right depending on the direction of the magnetic field, north or south. Even though the particles are going fast with a lot of energy, the magnetic field penetrates far out into the solar wind. A little deflection is enough to force most to miss the entire Earth, creating a charged particle shield.

If it doesn't miss, it gets trapped by spiraling around the stronger magnetic field lines closer to Earth. This sends the slowed particles into the magnetic poles, creating the amazing glowing displays known as auroras. Fast electrons and slow protons collide with atmospheric atoms eighty kilometers up. This excites the atoms, making them reradiate energy as a characteristic nitrogen-colored glow.

This planetary solar firewall isn't just an option for life; it's essential. This is the main reason Mars has no ocean and no significant life after four billion years. It cooled its smaller core early, stopping it from rotating and losing any magnetic field it might have had in the beginning.

Our planet is alive in a way. It is always changing because of thermodynamics and other natural processes. These processes happen at temperatures that greatly affect liquid water chemistry, from ice to steam. Water creates an ideal setting for organic chemistry with thermal and solvent abilities producing stable and temperate conditions.

Between 1.5 and 2.5 billion years ago, Earth, oceans, and the atmosphere stabilized. This stability allowed organic chemistry to thrive, supporting carbon complexity. Soon, tiny worlds of organic ecology began to form in warm, shallow tidal pools, creating ideal habitats.

The moon plays a key role in creating tides. These tides cause warm waters to ebb and flow, bathing the organic soup twice each day. It's like a huge global hydroponics tank. It has sat in the moderate sun, getting stirred by the moon for millions of years. With that much time, something is bound to grow, likely scummy and green. For millions of years, amino acids learned how to form giant protein molecules that can move and replicate themselves.

Sunlight is vital. Some molecules harness its energy to break chemical bonds. This process enables other reactions, which might seem unlikely, to occur and thrive in the mix. A special molecule called chlorophyll can convert solar energy into chemical energy. It does this in its own unique cell, known as a chloroplast. This cell works in conjunction with other cells. It uses photons and combines carbon dioxide with water. This process makes glucose and oxygen. Glucose and oxygen is a key source of energy for other life forms.

Glucose is burned for energy or turned into other molecules. Meanwhile, oxygen is released into the atmosphere. It pollutes Earth's early atmosphere. This atmosphere mainly had carbon dioxide, methane, ammonia, water vapor, and hydrogen. Life adds oxygen to our atmosphere to the degree it can now support open flames and oxidize metals rapidly. Large swaths of land turn red from the rusting of iron in the ocean.

Our atmosphere is made by life, for life, and is in a constant delicate balance, challenging life's ultimate goal for survival. From the start, cells show a clear separation of roles. Some living systems support others by natural sacrifice. This helps life grow in complexity, eventually leading to awareness.

When we study this closely, we see it's a matrix relationship. One life form doesn't just support another; they all work together. This teamwork helps the whole system function as one. It's not a single linear function, but a matrix of functions. If one piece of the intricate puzzle disappears, the whole thing falls apart. We don't have to eat animals, but we have to eat organic chemicals.

Algae is the bottom food source of life. It thrives in sunlight, CO_2, and water. They multiply quickly and store energy. They also create basic organic molecules that other life forms need. These forms consume and process these molecules, returning elements to the algae. This cycle helps keep the basic food chain going endlessly.

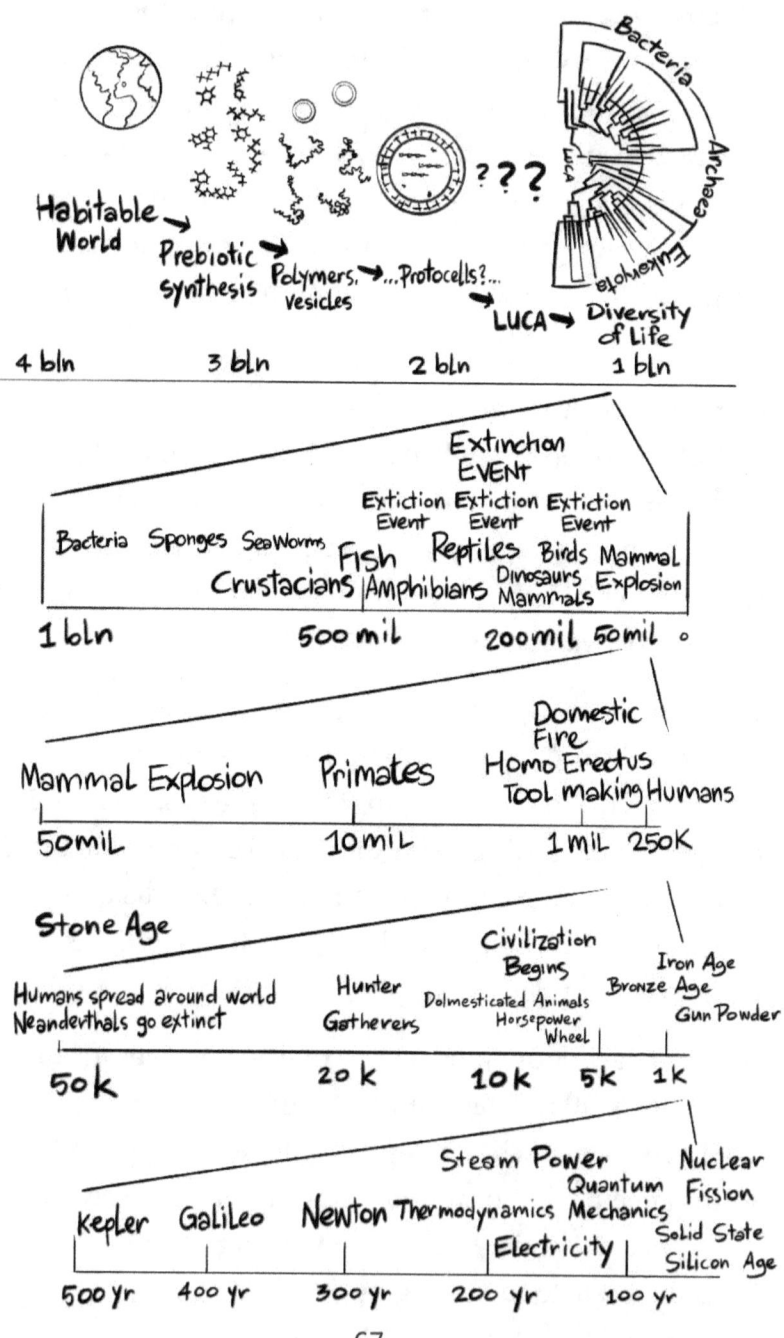

In nature, motion mainly happens because of temperature or height differences. Molecules tend to move from hot areas to cold ones and from mountains to the sea. Complex molecules require energy to move and create new ones. They use special energy molecules, like glucose, to transport energy itself.

Energy is needed for building or duplicating large protein structures. Cells begin to cluster in ecosystems. This creates symbiotic relationships that help them survive and grow in number. Some learn to live side by side while others learn to live inside one another as independent contractors to life's processes.

From the very beginning, life has but one motivating goal: to survive by any means possible. It replicates itself and plays the big numbers game using large number probability to ensure ultimate success. It will try any configuration, test every possibility, bend over backwards to conform, in order to make sure its kind finds a path to survival. Large number probability becomes a sound strategy for almost all. Spread your seed around by the millions and somewhere it will find purchase.

The figure above shows what happens after the solar system is about a billion years old. By then, the ocean and atmosphere are stable and temperate. Next, organic amino acid chemistry takes another two and a half billion years to evolve into proto-cells. This process moves Earth from one side of the upper diagram to the other. Once cells form, life begins.

This means life has created an encased chemical environment protecting its life-giving processes. It can manage this small space precisely. The goal is to turn chemical energy and amino acids into duplicates of itself so the information of its existence survives to make more. It's alive!

Today, we consider certain RNA molecules and viruses capable of doing only this. I consider them alive in every sense of what alive means, but I doubt if they have any feelings about it, one way or the other. They just are, as the rest of the observable universe seems to be all about. A singular exception to the rest of the universe appears to be us humans. We seem to be the only ones around who give a damn.

From this view, life might have settled into a steady balance between available energy and growing in numbers. It could have lived as long as the hospitable environment allows. But equilibrium doesn't last long. They may face random energetic events. These include being hit by an ionizing particle or photon. They might also get oxidized by acids or cooked by lava. Nature, however, is not interested in equilibrium when energy is plentiful, available, and flowing.

Cells take a big leap when they join forces with specialized cells. This helps them adapt to their environment and improve survival. They don't all act alone. Some learn to work as specialized groups that provide food and security for all.

Life forms that join together find survival in marginal environments much easier. This is a basic law. It says that one entity can survive better when cooperating with others. Their survival can increase quickly when they team up. There is an innate drive for cellular socialization as an organic survival tactic. It's buried deep in all of life's expressions.

Once multicellular ocean life emerged around three and a half billion years ago, it quickly grew in size and complexity. This period saw the development of nearly every life form and feature we know today. That is, until life became so pervasive that it changed the Earth's temperature balance. Life nearly ended when the Earth entered a runaway cooling phase. Global temperatures dropped sharply, causing ice to spread across the planet. This ice reflected more sunlight, leading to even more cooling in a cycle known as an ice-over event.

In the past billion years, life on Earth has faced four mass extinctions. During these events, over 70% of all species were lost. These events change radically the evolutionary path. They allow new species and animal groups to thrive in the new, often permanently altered conditions. During the ice-over event, life persevered in the form of algae. These algae can grow in ice and produce oxygen. This process helps warm the Earth back to normal. The Earth is as dependent on life now as life is on it.

And be certain that if a small asteroid hadn't hit Earth about 65 million years ago, mammals might not have evolved into many forms, including humans. The Earth can get stuck in stable organic conditions. This happens when a tight ecology forms. Sometimes, a major extinction event is needed to shake things up. Then, other life forms can have a chance to become the fittest.

For instance, sexual reproduction became the in thing to do some 450 million years ago by some nasty little fish. On the other hand, human feet appeared in the form we see today only about 2 million years ago. Our ancestors climbed down from the trees and learned to walk. They didn't stop until they spread across the world in just a few hundred thousand years. Our sex drive is far older and more ingrained than walking to the river for a cool drink of water. Who could have guessed that?

Our vertebrae with ribs enclosing our internal organs go back a little longer than sex. Our brain, however, goes right back to the very beginning of complex multicellular life forms. Specialized electrically active cells start to differentiate in a cooperative environment. Then, a new type of cell appears. This cell can generate and transport electrical signals through chemical reactions and ion drift. It moves these signals along cell walls over long distances. Its main roles are to coordinate, synchronize, and stimulate other cells.

It may, in fact, have another purpose before being put into actual long-term use, and that is to guide the growing stem cells in an embryo. They may help the blastomeres with RNA unfolding and orientation. Proteins guide this process, showing where specialized cells should go in the structure. Once extended, it leaves the ends of its dendrites in position to act as a communicating nerve cell. It shifts to sending and receiving signals from the brain like a permanent local government agent planted in our midst right from the start.

Once in place, they wait for the right conditions to fire their characteristic electrical signals. Signals move along the dendrites to the cell nucleus. There, they transfer to other dendrites. The strength of this transfer varies based on past activity. It learns and remembers by imprinting paths.

Its role in the survival game is to coordinate motion cells. They activate when a specific electrical sensory condition is detected. The most basic operation may have been detecting sunlight. A light-sensitive neuron sends a signal to a small group of interconnected neurons. These neurons then activate, sending small voltages signals to electrically sensitive cells. When stimulated, these cells contract or relax.

This makes the whole structure move. It depends on whether it learns to go toward the light for easy food or to the dark for protection from others. They begin with random assignment. Those who take the wrong turn don't survive. But the lucky ones do. They pass their luck to the next generation as a hardwired solution. Soon, the cross-connect in the middle is set up with the right learning thresholds at birth, giving it a head start in the cellular survival game.

In this chapter, I've provided a broad view of our universe's history leading to life. This helps the reader see its vastness, understand how it works, and appreciate its uniqueness. We need to understand the evolutionary paths that bring us inevitably to this moment. The Earth may be facing another mega extinction event. There's a clear exponential rise in advanced thinking and the manipulation of life forms. This trend culminates in our fast-paced technological rush toward the singularity.

The big break for this acceleration happens when humans start practicing science. They aim to maximize true knowledge and discover one of the universe's biggest forces: electromagnetism. Maxwell explains it fully with his four simple equations. This helps us discover quantum mechanics, relativity, semiconductors, microprocessors, and now, artificial intelligence.

Another big step forward happened around the same time as the periodic table and Maxwells equations. Quantum mechanics came into being helping us understand the nature of the chemical bond. With these two major achievements, humans burst into the modern world of science and technology. They made a nuclear bomb impact and have never looked back.

Chapter 3: How and why life progresses.

One of the things that still makes scientists shake their heads and ponder is the apparent violation of one of nature's most prevalent rules. When left alone, energy flows from an organized state to a disorganized state. In fact, the game of modern economic life is to generate energy, and on its way to becoming disorganized, we divert some of it to make things move to our will. It's how we use technology to gain power over inanimate objects. We can dig big ditches and pile up big rocks at the expense of Earth's hydrocarbon storehouse.

It's how we build artifacts, move ourselves, objects, and improve our ability to control and affect our environment. Hominids first got energy by burning carboniferous plants or wood. Wood contains a large amount of hydrocarbons, primarily in the form of cellulose, a polymer that makes up the skin of plant cells. When heated, hydrogen disconnects from the carbon and combines with oxygen to form water vapor, H_2O. The leftover carbon combines with oxygen to produce carbon dioxide, CO_2. These molecules fly away from the burning material at high speeds. This happens because heat is released when hydrogen atoms and carbon atoms drop into a lower energy state. They jump from the loosely bonded carbon chain to a tightly bonded oxygen releasing more heat energy.

The energy goes into kinetic energy, the energy of moving masses. $E=1/2\ mv^2$ defines the kinetic energy as one-half the mass times the velocity squared. Look familiar? Remember Einstein's energy for mass equivalence, $E=mc^2$? This is its low-energy cousin: the non-relativistic equation for kinetic energy at speeds well below that of light. When two fast-moving objects collide, the impact can tear their structures apart, even down to the sub-atomic level.

We touched on this earlier when I mentioned that nowhere in nature do we see rocks rolling uphill on their own, but that is in fact what life does. It reverses part of the flow of organized energy to disorganized energy by producing work. This is opposite to nature's natural drift. In other words, life exists because of energy flows which organic chemistry takes advantage of. It finds a niche in the environment where another natural law can take hold. Organic chemistry survives by creating complexity and diversity rather than simplicity. In nature, the lower energy state is the more disorganized, containing the least information. The price of information is energy.

The idea of life as a low-energy fragile chemical process might seem strange. It happens naturally, even among much more energetic processes. It's very unlikely for an organic molecule to grow in complexity and become intelligent by itself. This is because, as soon as such a molecule forms, many processes start breaking it down or pulling it apart just as quickly.

Most chemical reactions are reversible if there is little or no energy difference between the two states. But if there is, then the system loses the excess energy as heat, and the whole process settles into a new state. At some point, you might think organic molecules will grow too large. Natural forces will then stop them from getting bigger or more complex. Why does life seem to violate physics?

The answer, of course, is that it doesn't. The universe is a closed system of matter, energy, space, and time. It is by definition consistent within itself; all physical laws work everywhere, and all events are reproducible. Time separates events and sets the rules for causality. This means that what happens first must cause what happens next. In other words, there are no more gods creating stuff out of thin air; whatever is here has been here from the beginning, and it's not going away anytime soon.

This summed up by the conservation law of energy and mass. You can convert back and forth between these two descriptions of energy, but you can't destroy it or create it. Other key physical conservation laws include CPT, charge conservation, parity conservation, and time conservation.

Charge conservation says that electric charges, such as leptons and quarks, are not lost. They cannot be created or destroyed, only moved around. This makes the whole world of electricity work the way it does by giving us the power of the circuit. We can force electrons from one place to another, carrying energy or symbolic information. By trapping electrons in circuits made of conductors or semiconductors, we can create electro-mechanical machines. These machines use electrically generated magnetic fields to move objects creating work.

Parity conservation is a special case of inverting only one of the three spatial dimensions. It's the universe as seen in a mirror. It reverses left for right. In nature, we most often observe that left and right are only relative and not absolute physical differences. But if you look at your hands, they are indeed very different objects.

For instance, you can't shake hands with the same hand. They always have to be of opposite configuration, like sex, to fit together properly. In nature, we do indeed find instances where some forces do not obey parity conservation. What happens in reality when viewed in a mirror simply cannot exist.

It turns out, nature has a slight bit of left-handedness going on for rare events at an extremely fundamental level. This doesn't concern us much, but many life forms, starting from cells, show a symmetrical structure. Their bodies are arranged where one side is a perfect mirror image of the other.

Organic nature has many left-right asymmetries due to the basic structures of folded protein molecules. One form of a molecule can take part in an important life chemical reaction while the other might act as a deterrent. In chemistry, isomers are molecules that share the same elemental formula. However, their atoms can be arranged differently creating multiple physical layouts.

The last concept is time conservation. It applies to life in a limited way. It means that for every action moving forward in time, the reverse can also happen. If two hydrogen atoms bond with one oxygen atom, they release energy as photons. An identical photon can then hit the water molecule, breaking it apart into individual atoms again.

In thermodynamics, a whole class of processes are reversible. This happens when they change energy states in small, steady steps. Each step can be reversed without losing information about the path taken. Steam engines and AC motors can run in either direction equally well. There is no preferred direction in time except for entropy.

If you open a small box full of air inside a larger box without air, the air molecules will spread out into the bigger box. They won't magically gather back into the small box later. However, no physical law stops this from happening. If we look at it as a series of probabilities, we see that the chance of all molecules ending up in the smaller box at once is nearly zero. You'd have to wait longer than the age of the universe to even see it happen, but it is possible.

In thermodynamics, non-reversible processes show a key tendency. Low entropy thermal energy tends to flow toward areas of high entropy. This means organized energy becomes disorganized with time. Organized energy is hot and disorganized energy is cold. In our case, energy containing information is hot and energy with low information is cold.

At a certain low temperature, all thermal energy stops. There's no information left, and everything is still. Nothing vibrates or moves. The law of entropy states that all kinetic energy will eventually fade. In the end, everything will rest just above absolute zero, stuck in its ground state forever.

So, life isn't really violating any physical laws, although it bends a few around a bit. Temporary setbacks for entropy are quite allowed, as long as you pay the price of always expending more energy than you get. It will take smart thinking, a lot of energy, and a lot of effort to fix our technology problems.

Another way to say this is that life starts with a debt of low entropy. This means it begins with a lot of new information, like fresh cells and clean growth instructions. Slowly, we lose this information until it is all gone by death. Overall, life pushes entropy lower while the rest of all natural processes push it higher. For cells to live forever, all they need is a continual supply of new information to replace that which is lost naturally.

Needless to say, the early beginnings of organic chemistry leading to life on Earth are still pretty much conjecture. Fossils provide little evidence. We guess some clear initial conditions and apply processes we know. We add in the right chemical equations over time and observe what happens. We have better information now by recreating or simulating early Earth conditions, and we can see how they might play out in the lab. But that's a far cry from what actually happens primarily due to time scales. We've known about chemistry for just a few hundred years. We understand how it works, but nature relies on trial and error. It takes time for large number probability to kick in making big changes.

Our best shot at this is to set up the appropriate physical conditions and rules of chemistry in a computer simulation. Then we can vary the conditions to see how it plays out over the time scales actually involved. This gives a pretty good guess on what happened but not much on why. All I can say about why is that life is basically a teen-bad making poor decisions.

It often does things for the simple reason that it's possible. If young people chose only what seems cool or doable, many more might win Darwin Awards each year. People who remove their genes from the gene pool through reckless actions are recipients of the Darwin Award. This award honors those who are basically stupid and remove their genes from the gene pool, improving the overall DNA of the population.

But nature doesn't care whether anything survives or not. What nature does is express probabilities and wait 'til something hits the jackpot. After all, if life is at all possible or probable, then after organic chemistry stews for a billion years or so, it becomes almost certain. The law of big numbers describes a universe that appears as a particle-wave duality. In this universe, the wave component is a wave of probability amplitudes. Existence is like a game of chance. Quantum mechanics through probability calculations governs the tiny atomic world. It shapes the chemical reality that makes up everything.

We must also think about the early stages of life's chemistry. If we found these conditions in isolated, sterile tidal pools today, they would not last. Any higher life forms would eat them, or advanced life would harm them with pollution. The birth of life on a planet can happen only once, if it is to happen at all.

All life on Earth comes from this planet. We share the same imprint of ancient chemicals found in tidal pools. These chemicals still flow in our cells today. There is no sign of anything from beyond Earth, like gods or aliens, occurring in the making of life. Everything we are or will ever be comes from right here.

A key property of a cell is keeping a balance of ions across its membranes. This balance helps draw chemicals inside using electrical forces through its semi-permeable wall. This maintains its internal chemistry lab working under the same conditions it developed in a long time ago.

At the molecular level, cell walls separate chemicals based on size and shape. They can also distinguish isomers, which are the same molecules with different shapes. A dynamic electro-chemical filter can, in some thermodynamic systems like life, defy entropy. It organizes energy while using very little energy itself. It can exhibit osmotic behavior where some molecules pass the membrane in one direction but not in the other. This is used by the cell to separate and concentrate chemicals necessary to carry out its normal energy-driven chemical and electrical processes.

The early oceans had a simple mix of sodium and potassium chloride salts, with a few trace minerals. A living cell needed to capture this environment. This way, its electrochemical makeup would mirror the ocean. The cell could focus on its work, unaffected by time. In this case, the oceans two and a half billion years ago had about a third of the salt it has today. All cell interiors today have the same salt content. This is true for all life forms. It comes from the early Earth's ocean, where all life chemistry began.

So, let's just say for now that natural forces are in play for the full billion years and no other explanation is required to end up with the first cells. It's a stretch of the imagination, but we are up to it. The key takeaway from the evolution of the first cells is that they began as simple single-celled bacteria.

These bacteria formed symbiotic relationships with other cells in the same environment. Symbiotic and parasitic relationships exist all through single-celled life. Early cells had functions like energy production and making life chemicals. They often lived close together, where one cell's waste became another cell's food. Pretty soon, they get rid of the middle man and just got married forever.

We eventually see a sudden explosion of various multicellular entities. Each has a unique internal chemistry, but they all share one key feature: the cell nucleus. Cells without a nucleus are like one-trick ponies. They are bound by the rules of chemistry to perform the same function over and over with only the ability to duplicate themselves as they are. If they are not damaged in some way, they can duplicate themselves by simply splitting into two identical pieces. This is called cloning.

A living cell with a nucleus has a special organelle inside it. This organelle acts as a chemical controller. It helps express the genetic codes found in RNA and DNA. This helps make proteins and enzymes. They control cell growth and division. They also record and store information for building these proteins. This information gets passed to the next generation. It creates duplicates under programmed control instead of only changing randomly.

If a bacterium improves its function, it can only share this benefit through direct cellular division. A nucleated cell can track key changes in its chemistry when needed. It then shares this new information with future generations, creating an improved version.

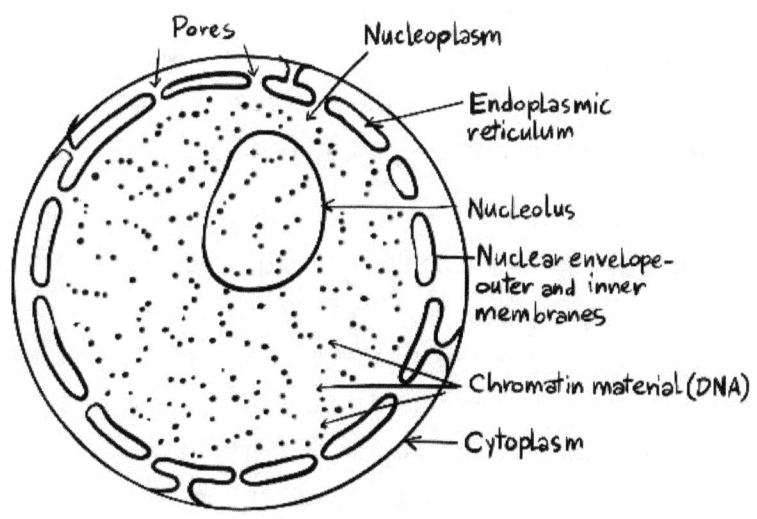

Bacteria and viruses have stayed pretty much the same in function and appearance. They haven't changed much since their early evolution, when they were the sole life forms on Earth. They did a lot of evolving to get to where they don't need to evolve anymore. They're stuck in a mindless chemical loop they can't break out of.

But you have to admit, the design seems to work pretty well since they are still around munching on us higher life forms like there's no tomorrow. But what the nucleated cell or eukaryotic cell accomplishes is a huge advantage on the road to higher life forms.

What it did was open up a huge variety of expressions of this basic configuration. DNA holds instructions in a sea of RNA, proteins, and enzymes. Ribosomes help move these delicate chains. They cut, slice, and dice to create essential protein structures needed for building a cell. Nature shifts from strict specialization to generalization. Now, specialization happens when needed, guided by an active program. DNA programs ribosomes, which are the 3D printers of proteins. Change just one instruction and the whole structure can be altered.

The odds of finding a cellular life form that suits its local environment skyrocket. Cellular life has changed the odds and now starts popping up in different versions everywhere it has any chance at all of surviving. When it gets a toehold, it can quickly change its programming. This helps it self-tune performance and adapt to the new environment. Evolution is now under the control of cells and their almost infinite adaptability and specialization.

At a critical point in this explosion of life, cells gain a new foothold in the organic soup they find themselves in. A cell consumes a smaller cell. But instead of breaking it down to collect its amino acids, they end up in a Mexican standoff, so to speak. By having the right kind of membrane that can stop the digestion, the smaller cell survives. The larger cell adapts from doing the normal messy food production and begins outsourcing it to the lesser cell.

This organelle called mitochondria is a cell within a cell with its own DNA and chemical factory. This chemical factory has one main job. It takes in sugars like glucose and some oxygen. Then, it produces adenosine triphosphate (ATP). ATP is the universal food for cells and their main energy source for all living functions.

Animal muscle cells have the most mitochondria for good reasons. Brain neurons also have many mitochondria. These mitochondria move along dendrites to deliver energy and calcium ions. This process activates the cell's electrical signals at the axons and synapses. It helps set up switching circuits and triggers thresholds. It plays a vital role in all cells' ability to live, and in some cases, the mitochondria may decide when the cell dies.

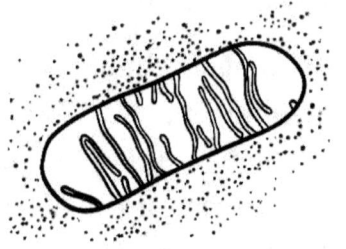

The cell now has its own small energy-generating plant that it can use independently. It has its own memory and specialized programming that allows for some independence of purpose. All future specialized cells get energy from this organelle, and in return, it gets a safe environment to live out its chemical purpose. Every cell now has its own mobile power plant allowing almost unlimited multicellular structures.

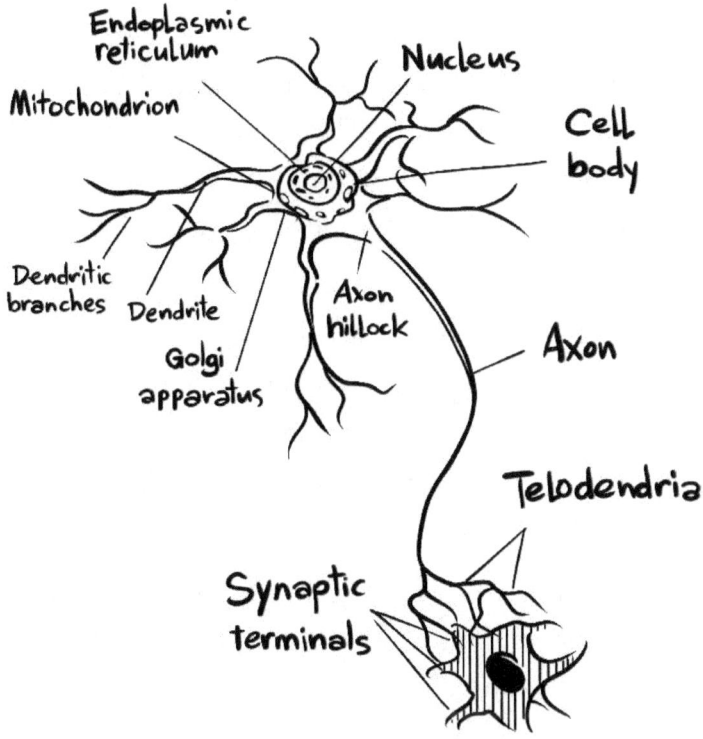

At this stage of multicellular development, some specialized cells gain electrical properties. They use chemical ions, mainly calcium, to send signals. Charged molecules have either an extra electron or one missing. This creates an electrical charge. External electric fields can then move these charges around. Neurons appear alongside normal structural and chemical cells. They coordinate the delivery of critical ions and chemicals across large cellular distances.

This controls how proteins unfold to read DNA and build the right amino acids in the right places.

Neurons are crucial for the growth and coordination of other cells. They help specialized cells work together like a single machine. They create the electrical fields that move proteins and enzymes. This guides where they should gather and form cell structures. This sets animals apart from plants, fungi, and single-celled organisms like bacteria and viruses.

The world of life has now split into three main branches. One is us, the animals. The other main group is plants. The third is fungi. Everything else falls into single-cell categories called eukaryotic. This includes algae. Algae protect living ecosystems and provide proteins to almost all life.

But we are on the track of finding out why humans are about to go extinct and do not even know it yet, so I'll restrict our discussion to animals.

As time continues to unfold on Earth, the animal kingdom begins showing off its diversity right from the beginning. Multicellular animals in the ocean have plenty of empty territory to expand into, and they take advantage of it. Around 650 million years ago, in the first 100 million years of life on Earth, animal body forms begin as simple tubes. They suck in nutrients at one end and expel waste at the other. They have a brain, of sorts, made up of many tangles of neurons all near each other, with some having axons going out to all parts of the simple body.

Neurons create a network that sends and receives electrochemical signals. They do this to either show a change in state or to indicate another cell's action. The point where they share multiple repetitive signals creates a decision-making neural network. This network can switch signals between paths. Sometimes, it even generates new signals from one-time chemical reactions.

I compare it to the old telephone networks before electronics. They consisted of wires linking telephones to a large switching board. This board connects phone circuits to nearby phones or to other switches, which link to more phones. It transfers information and is smart enough to know how to find the proper recipient of all the information it conveys.

Early computers were set up by linking wires across circuits. This connected the hardware to work in a specific way based on the arrangement of the logical parts. All networks have the ability to show signs of intelligence if they include some kind of intelligent control. For the telephone network, the operator is the intelligence. They connect calls, tear down calls and help spread rumors after chats.

A group of neurons acts as an information processor. Some neurons receive signals, while others send them out. Early neurons may recognize others of their species. They do this through basic smell, sensing unique organic molecules, or sight. They can detect light, focusing on specific colors, shapes, or textures.

The giant neural cell uses atomic-scale quantum mechanics to move energy and information. The trick occurs in assigning the proper response to any input. If a nerve cell detects light at its location, it signals the corresponding visual clusters. They drive resonant signals that create electrical patterns across similar clusters. The sum of the resonances creates a pattern of cross-connects. This sends signals to the right muscle groups in just the right amounts for necessary actions.

If it detects light, it might want to move toward it where maybe algae is turning sunlight into sugars. Or it might detect light and feel vulnerable, seeking instead to swim away to find a hiding place. The difference is in the learning. When a repeated action allows the cell to survive, it keeps it up. If not, nature plays the numbers game. It has many pawns to send out. This way, it can see who survives to multiply and who does not. Over time, this is the best way to create positive changes that boost overall survivability without design.

From here, a steady rise in complexity and diversity begins. Eventually, the Earth fills with animals, each having unique cells and structures. These features help them interact with their surroundings and ensure a stable flow of offspring. Neural networks are becoming more complex.

Neurons now specialize into over a dozen types. This creates a basic brain that can make simple decisions. It learns to react to its environment. Under survival pressure, it can adapt with different setups. This helps it trigger and switch neurons in new ways that can be passed on by horizontal gene drift.

The key details of animal evolution during this time are clear from fossil records. This means much of what we know is based on facts, not guesses or models. Life marks its presence in geology. Humans have studied this since the days of Greek philosophers who noted aquatic fossils found on land. They were often overlooked because they clashed with the prevailing beliefs. Then, someone realized they weren't proof of monsters from hell or rocks created for our entertainment. These are the remnants of actual living beings. They lived and died, leaving mineralized traces of their bodies. We will debate and ponder them for as long as we remain curious but the evidence is there and decipherable.

The fossil record shows us the evolution of life over time. We can see its ups and downs, revealing how much can change. And they do. We find evidence of extinction events or periods where the pace of evolution stalls or takes a radical new direction. This is plotted in the figure below, where there is evidence of at least five major events and many minor ones during our evolution here on Earth.

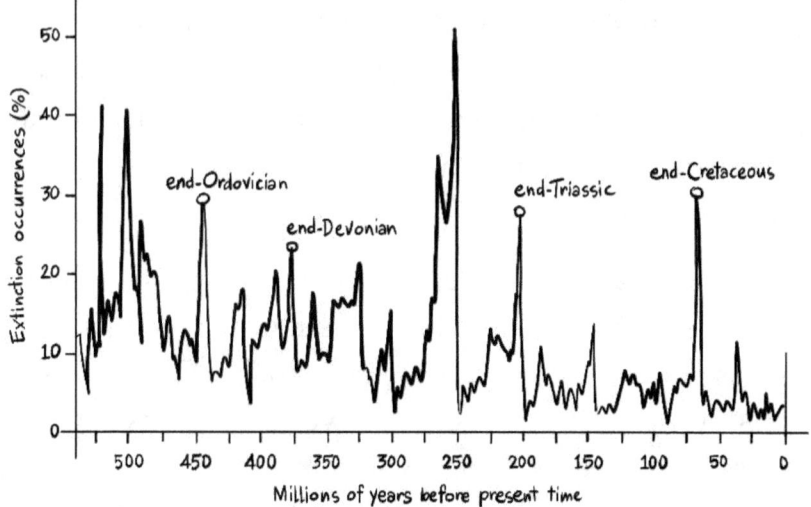

Most of these events, especially all the minor ones, are due to fluctuations in Earth's geologic environment. While animals are trying to make their way in the world, the Earth is grinding away at its own destiny. This can include some pretty violent and catastrophic events that are highly disruptive to life. The end of the Ordovician period is known for producing many cheap trilobite fossils today. These fossils are plentiful in curio shops and are about 450 million years old. Their numbers reveal just how widespread and long-lasting trilobites were.

The Earth became active for a period, causing the chemical balance of the atmosphere to change, causing the Earth's energy budget to change. Trilobites were small ocean creatures. They had three body segments and two light-sensitive domes on their heads. They were about the size of a mouse. Life was mainly sea animals with protective shells.

They scurried about coral reefs, creating a vibrant environment.

The Earth became volcanic. It released a huge amount of carbon dioxide and hydrogen sulfide into the air. This caused global cooling that disrupted shallow sea environments. Almost 85% of all life was killed off. Algae and cyanobacteria, at the food chain's base, were affected too. This led to a drop in ocean oxygen levels. At one point, the concentration fell below survivable levels. This triggered a rapid anoxia event, suffocating all higher animals.

Researchers found that global climate change first warmed the planet. This caused algae to bloom wildly. Then a cooling period began, leading to a glacial period on land. In this time, cyanobacteria consumed all the oxygen due to their rapid growth. When these conditions change and the planet warms, algae blooms occur. This leads to a bacterial explosion, which stresses the ocean's ecosystems even further.

Today, human actions are driving global warming. We release more greenhouse gases faster than during any past extinction events. As a consequence of all life being confined to the ocean at this time, many of the effects of glaciation on land had fewer effects underwater.

Not so during the Permian extinction, which happened around two hundred million years later. At that time, the land was filled with animals, and the sea was teeming with fish, amphibians, and reptiles. Vertebrates showed up with a body structure that worked well in later generations. Lungs, internal organs, and limbs help with movement. Some have fins or feet with five phalanges.

A large brain sits at the front, close to the eyes, ears, and nose. It processes sensory information in real time. The brain can now evolve its programming by learning and passing on that new neural configuration in the form of modified DNA. Things are popping until the Earth belches and the biggest extinction event we know takes place, almost wiping the slate clean.

Big eruptions in Siberia let out a lot of carbon dioxide, sulfur dioxide, halogens, and metals into the air and ocean. This causes global warming, acid rain, oceanic acidification, ozone reduction, and metal poisoning. There is a tremendous loss of biodiversity where 81–94% of marine species and 70% of terrestrial vertebrate families go extinct.

Almost all life on Earth is harmed. This includes trees, plants, lizards, proto-mammals, insects, fish, mollusks, and microbes. Earth's ecosystems settle down after roughly two million years. After that, life begins to grow more complex and diverse as a result of changing conditions. One lineage of survivors produces our ancestors, the first mammals. Dinosaurs appear and the famous Jurassic period is upon us.

The last major extinction wasn't the Earth's fault at all unless there's fault in being in the wrong place at precisely the right time. For early Earth, the solar system was crowded with debris. Celestial objects often collided with one another. Just look at any planet or moon that lacks atmospheric erosion, and you can see this violent past with lots and lots of craters recorded over eons. Here we are, some four billion years later, and most of the rocks that are going to hit something did and are removed from the threat pool. By now, asteroid hits of a size large enough to cause some real damages are extremely rare. But Earth gets hit anyway, and it's a damned good thing for us that it did.

After every major extinction, there is a period of recovery where the ecosystems are in flux and mutations have to keep up with new survival issues. The asteroid that wiped out the dinosaurs also ended their rule. They had dominated Earth's ecosystems, controlling resources for over 200 million years. This allowed them to thrive while mammals struggled to survive as small, underground creatures. After the big dinosaurs disappeared, the smaller creatures got their chance. They quickly moved into the new, open, but empty ecosystems that were left vacant.

Mammals are now thriving. They are spreading into new ecosystems and taking charge. Their better reproduction methods help, along with their smarter brains. They adapt well to the unstable conditions during the glacial periods.

The earth's crust is moving. This movement raises volcanic activity. The Atlantic Ocean forms over a new crack in the earth. This crack splits the giant continent Pangaea in two. The earth goes through multiple global warmings and cooling. Mammals spread to every corner of the earth, adapting themselves to whatever they find. Later periods saw mammals adapt even better, leading to giant versions similar to dinosaurs for reptiles. However, unlike dinosaurs, giant mammals did not dominate their environment for long. They have to wait for the primates to appear to cause that much disruption.

The Earth is active due to continental drift. This movement pushes continents around. Some species get stranded, leading to different mutations. These mutations have varying chances of survival, shaping their own unique futures.

It's a chaotic mix of shifting ecology. Ice moves forward and then retreats. Global temperatures and extreme weather patterns change constantly. Volcanic activity from continental movements releases gases and pollutants into Earth's thin atmosphere. This alters global chemical and energy balances. It creates a natural setting for new species to emerge.

Out of this turmoil, you're bound to get something like the flora and fauna of Australia and nearby islands. Being separated from other species and having a warm, stable climate helps marsupials thrive. In the variations of how sexual reproduction is carried out by the body, the choices are limited. You can hatch them outside the body, inside the body, or use a mix of both. In the hybrid method, the fetus moves halfway through its development. It shifts from an umbilical cord to a pouch that produces milk before it is fully born.

The marsupial birth is probably one of the least traumatic ways to come into this world. Think about that pouch, always being there for you as you test the outside and develop a slow taste for living free. Humans are unique. They mimic this by holding and carrying their babies by hand. Babies can be swaddled or cozily placed in a pseudo-pouch or backpack while being fed by their mothers. Don't be surprised if, when the aliens finally show themselves, they turn out to be more like efficient marsupials then mammals.

About 5 million years ago, Africa was full of primates. Thick jungles covered most of the continent during this time of intense volcanic activity. Global warming is drying out many areas in Africa. This is especially true in the central rift valley. Here, small volcanoes erupt, sending pumice into the air. These eruptions push the land apart. As a result, the fertile valley expands and grows wider. The warming causes the vegetation to thin into mostly grasses, and the jungle turns into a savanna.

During this stressful time, some tree-dwelling primates came down from the trees. They learned, or forced themselves, to walk upright. They moved between food sites, watering holes, and safe areas. Before, they stayed locally in thick vegetation, feeling secure and free from stress. During the time of big predators, the extra stress on this strain of primates must have been huge. Survival must have been an overwhelming challenge.

The primates respond by mimicking their natural enemy. They start using new resources, shifting their diet from fruits and vegetables to meat and tubers. A high intake of protein and calories leads to many changes in the body. They developed bigger muscles for strength, improved feet and hips for running, and stronger upper body muscles for throwing objects. Muscular hips develop, and a wider pelvis in females helps support longer pregnancies.

Most importantly, there is an increase in brain matter. They find that cooking meat with fire has many benefits. It keeps the meat fresh for later, makes it easier to digest, and boosts its usable protein and fat content. This is key for a body and brain that requires a high metabolism.

Isolation over time leads to gene variations. This causes DNA mutations that push species into new ecosystems. They extract the last available resources, which is key for survival. But with the bigger brain comes more than just thinking about where your next meal is coming from. It lets them watch, record, and replay the natural activities they witness daily. They can mimic, play, or interact with the events until they gain knowledge. Here, taming fire gives these hominins a big edge over predators. It also puts them on a definite path to technology and eventually becoming humans.

They had been using sharp rocks and digging sticks to do a lot of their processing of food. Now, they make sharp rocks by banging one against another just the right way. The result is knowledge. They bind twigs and stems into rope, which can be used to build snares. That's pure knowledge. They carefully tear leaves into strips, which are woven into baskets, slings, and clothing. Again, something to be learned and passed on as a key advantage in their struggle for survival.

Hunters process skins for clothing and create all sorts of ingenious things from their kills, making life a little more tolerable. After all, the environment is going crazy as hominins expand out of Africa into new lands. They bring one thing with them that makes them masters of whatever domain they might land in.

Their brains and its technology provide the key advantage. They dominate Africa, Asia, and Europe. They look for new habitats where it's warm enough for game and plants. The Earth cycles between warm and glacial periods. This shifts local environments in the northern and southern hemispheres. It causes almost continual human migrations.

This is the case for our closest ancestors in Africa at the time. Heidelbergensis lived near the Rift Valley and the Nile River in Africa. This was about four million years ago. They gave rise to various hominin species across Asia. Then, around two hundred thousand years ago, our species, Homo sapiens, emerged as modern humans with a unique DNA. Neanderthals, for instance, are genetically cousins, not ancestors.

Paleontologists are starting to discover new hominin fossils around the world. These fossils show the many mutations of the successful Heidelbergensis design. We owe them a lot as they got us started with technology and the use of information to gain virtual knowledge.

They are very successful and productive. They creatively use available natural resources efficiently and thrive in a competitive environment. They create many successful variations of their basic design and function. They keep making changes slowly until we show up and seemingly break the mold.

The key takeaway from the journey of carbon-based life forms to humans is the amazing solution we represent to the survival challenge. Our solution is simple. It's called virtual networking. It becomes crucial for our survival. We wiped out many of our hominin cousins and drove most of the Pleistocene megafauna to extinction. We conquered every habitable environment, bending nature to our needs wherever we went. Humans conquer the world not one at a time but with a whole mess of them all at once. And we don't play fair.

Heidelbergensis circa 50,000 yrs ago

Chapter 4: Where did our brain come from and where is it going?

The animal world has many types of neural networks. These networks help coordinate functions and organize control. The simplest nervous systems without brains are like the kind of diffuse neural networks in jellyfish and hydras. They might be an active electrochemical neuron cell mesh network, but their processing power is really low. This is due to their small numbers. Diffuse neural mesh networks work well for guiding a blind body. However, they struggle to combine information and take life-saving actions.

Neural networks help flatworms move and guide butterflies and bees. They provide various strategies for survival. They improve success by being more aware of their surroundings and quick to act in response to common threats. As mesh networks grow in size and complexity, they help guide animals to be more prolific at finding mates and more adept at working with others in symbiotic modes. This increases their overall fitness for survival.

They account for the complex songs of birds and whales and the creative and symbolic minds of humans. Above all, they are neither state machines like digital computers nor imitative like analog computers. They work by simulating reality, not imitating it. They can theoretically model anything. This includes state machines, analog systems like slime mold, and even complex vector machines such as quantum computers.

The brain is a gigantic communication network. It has different areas that perform special functions. It shows how species vary and how they connect with their surroundings, just like other body parts.

With advanced brains, animals can face challenges with an extra advantage: that of memory and learned strategy. Every boost in brain hardware leads to a big shift in survival skills. This shift decides which species can evolve and what type of body they will need afterward. So where did this thinking thing come from that now drives evolution?

The neural network that becomes a brain is essential for solving early life mobility issues. This is especially true in ancient tidal pools occupied by the first multicellular living entities. It solves this problem by using input sensory data. It connects this data dynamically to choose the right pre-programmed output actions. Nerve electrical conduction isn't just about electrons as in copper-wired circuits. It involves organic molecules with extra charges dangling off them precariously and moving about by many subtle forces.

These molecules help neurons create brief electric fields. This process sends chemical signals from one end of a neuron to the other. Electro-chemical molecules do this with help from special cells. These cells generate, receive, and control electro-chemical signals, pushing small pulses of electricity along hydrocarbon chains.

Neurons help animals sense their surroundings. They use this information to take the right actions. The brain is an information-to-knowledge-to-action processing machine.

It can be as simple as light being present or absent. Also, an active chemical, like an organic molecule, may indicate that

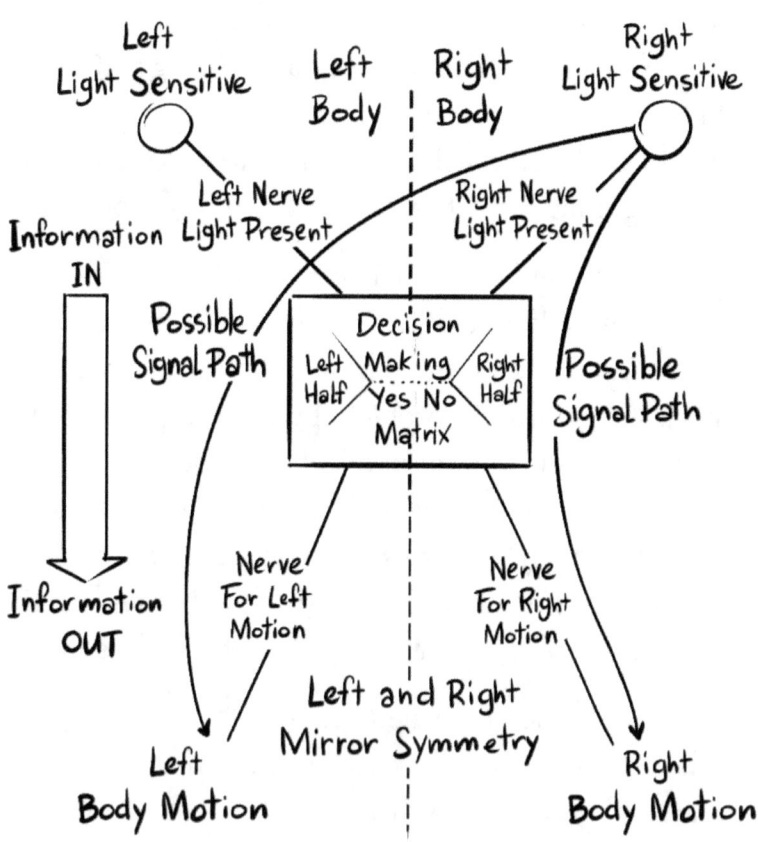

something alive is close by. To survive, it likely detects other animals of its kind. This helps them find resources together or locate mates for reproduction. Only mobile animals, like worms, need these skills. Complex neural networks only form in creatures that actively search for food. They don't develop in those just sitting on a log, waiting for a meal to come by.

The two different attacks on life face opposite challenges. One must be very lucky to be in the right place at the right time. This luck needs to last long enough to find food and reproduce. Its plan for survival is strictly based on numbers. Throw enough seeds out there, and even the most improbable locations will be reached and tested for fitness.

Survival of the fittest becomes survival of the luckiest. It works pretty damned well. Look at the diversity of plant life all over the planet and in places no animal could ever go. The Earth is a living planet that lives by the rules of nature. Large number probability is a key element used by nature for survival.

Mobile life is in a constant state of distress, forced to seek through controlled locomotion, food, sex, and security. Distress measures how active the brain's electrical signals are. When these signals speed up in numbers, distress increases. The body needs more resources, such as hormones, proteins, amino acids, and oxygen. Where there is decision-making, there is virtual internal chemical and electrical conflict.

The animal is in a state of constant quandary: does it stay or does it go? Does it stay where it is safe and food might continue to be available? Or does it move to escape trouble or to test luck and find more food elsewhere? *'Should I stay or should I go?'* is the existential challenge for all brains, and they spend a lot of energy grinding on this unending quandary. The first goal of any brain is to come up with a quick and dirty solution to survival, thus saving vital time, energy and stress.

When the nervous system solves this problem, it improves survivability. Fortunately, the neural network is built from scratch to solve this one dilemma. A multicellular body needs to be coordinated by some means in order to actively move as a single entity. Something has to tell all those movement cells when exactly and how strong to pull on the oars. They need a drumbeat and some persuasion in the form of the whip. For muscle cells, electric shocks do the whipping

The switch from a cross-connect decision-making matrix to a full-mesh neural network occurs quickly. With this comes a primitive mesh-networked brain. It has the capacity to combine information from different senses into preprogrammed responses. This ensures the survival of the quickest and the smartest.

The brain also helps create new body types. These new bodies have special limbs for movement and unique sensory structures to guide them. More complex brains evolve with

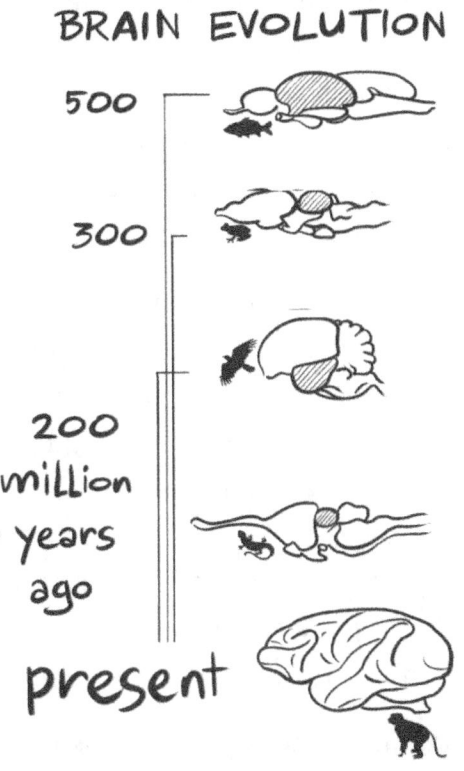

BRAIN EVOLUTION

500

300

200 million years ago

present

time, but most original models are still on the road and doing well. The last model on the bottom shows modern primates. It marks the final step in brain evolution. This human version is fully programmable and is connected to an external virtual network of similar brains. It represents a new type of brain that hasn't yet been fully tested for long-term survival. It's only been around for a couple of hundred thousand years. Compared to birds, which have a couple of hundred-million-year design history that included an asteroid, we still have a long way to go to prove ourselves.

We have learned that neurons do not operate alone in a sea of identical neurons. There are many types, with nearly endless cross-connection options. If the nodes in a complex network actively process information, they act as a computing machine. In an organic brain, it becomes a decision-making machine for fitness.

Neurons have a unique ability. They can develop into a specialized mesh network and process information to make decisions. There are many ways to achieve this. It often depends on the number and strength of nearby connected nodes. When these number in the tens of thousands and hey are tightly bundled into a matrix, they form a specialized processing cluster. These clusters, or neural nuclei, respond to specific electrical activity and produce a specific response. There are many types of neural nuclei.

The diagram below shows some typical network configurations for handling communications. Neural cells are arranged in a densely populated free-form matrix. Strongly symmetric networks are rare.

The ad hoc random-looking mesh network is the most common topology of the brain, although all types can be found. There are dendritic bundles, or neurons clustered together at their apical ends like a star network. There are columns of clustered cell bodies arranged in open spaces in the brain, making a hybrid ring and bus network structure. Column clusters vary in size and complexity. They connect diverse areas of the brain. This helps form large scale cluster

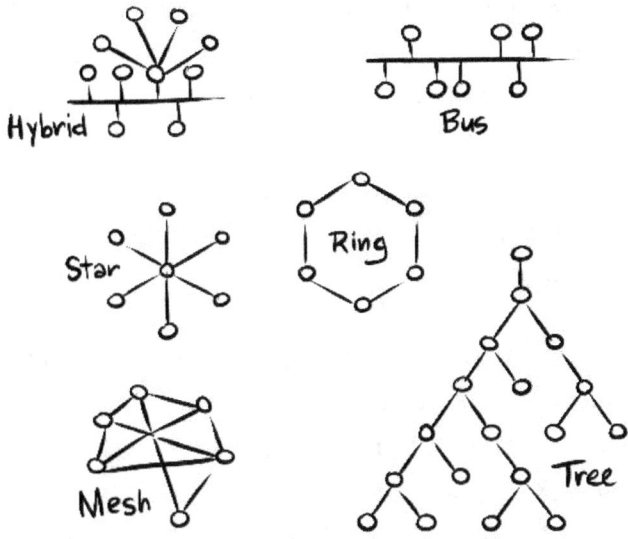

configurations responsible for awareness and thinking. They can even create temporary networks for exception processing. This lets them synthesize abstract patterns for learning. As a result, they gain coordination skills like playing a musical instrument or flying an airplane.

Neural nuclei generally have short dendrites. This helps them form close connections for quick processing of specific information, like visual or aural. They work well under set programming conditions.

Long axions with fixed firing levels send signals to distant synapses. This helps transfer information across large parts of the brain. They come grouped into fundamental but specialized resonating signal processor units. Many hundreds to thousands of neurons connect in a dense local network. This network is shaped both by genetics and traumatic experiences. It can be active at different levels or states of awareness at once, much like complex-vector quantum computers. They are technically state-representation machines for use in running reality simulations.

If they are located in the spine, they are called ganglia. Ganglia are switching matrices found in key parts of the body's peripheral nervous system. They play a role in autonomic processing. This is the old knee-jerk test of comedy doctor movies. Highly trained athletes practice specific muscular tasks repeatedly to program them. They call it *'being in the groove.'*

The brain nuclei play a key role in decision-making for mammals. The brain helps species use nature effectively. It also allows them to change their abilities without the slow and uncertain process of natural mutations. It learns and remembers. Through natural selection, it passes this knowledge to new generations. Now, these generations are better armored for survival than the last.

Neuron clusters often show synchrony. This means their signals are time-correlated. The brain doesn't have an internal clock like digital computers. Instead, it relies on resonant frequencies. These frequencies cause big changes in nuclei configurations for various tasks. Resonant frequencies create a firing rhythm. This helps brain-wide neural clusters work together. They avoid stepping on each other's toes while thinking. They cross-connect in an unplanned but meaningful manner, which creates whole new thoughts.

Resonance coordinates neural activity and provides an internal reference clock. This clock acts like a processing speed, much like the speed of microprocessors. It determines how much bandwidth and speeds the neural cluster can manage, maximizing its throughput. Along with the total density of clusters, these resonances constitute the brain's basic computing power.

It also must accommodate the time between signals sent out to the body with echoed returns so they stay in step and don't interfere. Most of the brain's core provides the electrochemical support for constant electrical activity. This is where the thalamus and pituitary are found. The neural clustered brain and its pulsating meshed networks step in unison maximizing throughput. They form a single decision-making machine. Its main goal is to provide survival skills for the individual.

Clusters behave consistently. They are controlled by fixed firmware, so they always produce the same output. This happens no matter what the stimulus source is. Input neurons can differ in their responses, so clusters rely on a best guess approach. They need to provide answers, even if the input signal has errors or inconsistencies.

The brain is deterministic because it provides a best answer for each problem. It is also flexible, as it seeks the best match to expected results. In the world of brains, close enough is usually good enough to secure survival. In fact, that's how it principally learns. A near miss is still a good outcome for survival. The trauma from this experience creates strong emotions. These emotions help the brain react better in similar situations later by only triggering those specific feelings. It learns.

Clusters are closely linked to how information moves and is processed. This can be represented using abstract network theory. A network of neurons links clusters. This setup helps the computing machine take in information, turn it into knowledge, and store it as emotion-tagged memories. It's a multi-state machine that can store and process information at the same time. Different combinations of ordered states make it unique.

When we remember, we piece memories back together from the stimuli. This process mimics how we first felt and sensed those moments. The brain always has incomplete information. It needs to reassemble this data into a reality simulation. Only then can we recognize and use it. In terms of computers, it is always overwriting the memory every time it is recalled. Again, to the survival brain, close is close enough, and it saves on memory space and the energy supporting it.

Neuron clusters are the key parts of thinking and other complex brain activities. They operate like a mix of firmware and software using electrical and chemical signals. The firmware includes the physical neuron connections. It also involves the non-linear chemical processes at synapses and the signals that travel along axons from one nerve to many.

The software keeps using specific synaptic conditions that makes it remember or switch to a lower voltage path. As a result, its normal response changes. It has altered its basic signal-firing thresholds, leading to a programmed response. Basically, the advanced brain can change its programming on the fly.

Neuron clusters wire themselves in patterns based on their activity. They also rely on electrochemicals to control triggering and nearby synapse responses. Neuron clusters usually respond constantly with repeating electrical signals in real time. Some produce complex chemicals, enzymes, and hormones. These can change the electrochemical balance at synapses and dendrites. This causes long-term changes in programming.

The simplest brain, what is termed the first transition in brain evolution, is not technically a brain. These basic nerve cells are part of multicellular structures. They start the process of fully integrating nerve cells into the structure's

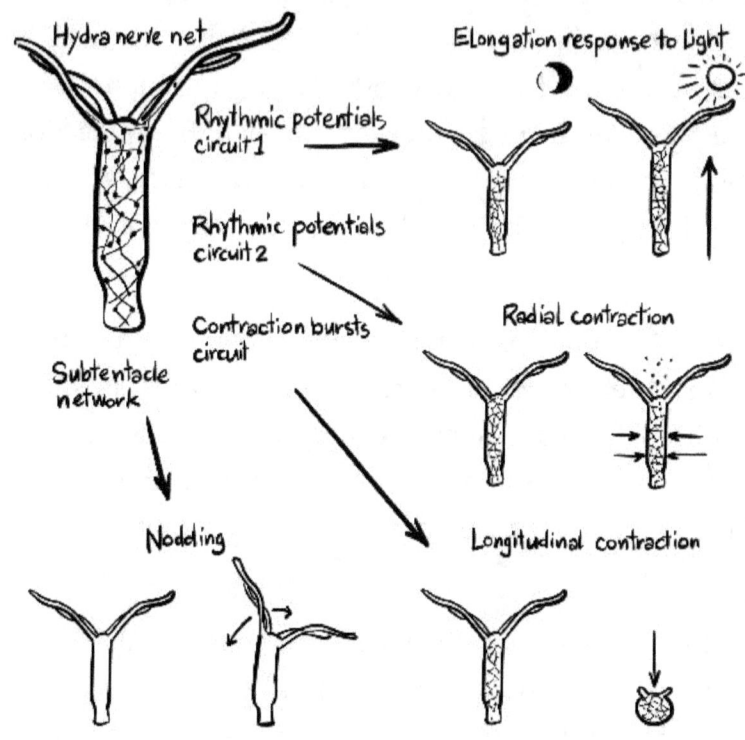

fabric. Its use is limited to simple coordination in order to attain some kind of movement. The figure above shows how simple nervous systems limit movement. This basic setup still works well for hydras, sponges, and corals today.

Movement means finding food before it finds you, and the nerve cells provide this level of coordination. Actual movement is attained for whatever reason. It's safe to say some find it helpful; hence, survival breeds more. Some find it less helpful and eventually die off. Some find that sitting still is not such a bad strategy. They proceed evolving without complex electroactive nerve cells in a centralized configuration. A coral is truly brainless but reactive.

Mesh networks work well for species that stay put. They use limited movement to respond to food availability. They swing and sway in currents to gather more food. If there is a threat, they withdraw to safety. They don't need eyes or ears. They sense touch, temperature, and chemicals. This information helps the entire entity survive better. Large numbers are vital for survival. They help coral colonies exist as one unit, covering thousands of miles of ocean.

The second transition stage of brain development starts when a centralized processing network forms. This network connects nearly all body nerves that used to work in a point-to-point manner. Now they are point-to-multipoint and vice versa, forming the first specialized clusters. To better understand how the brain works, I've added a flowchart. It shows how the brain processes information, makes decisions and causes actions.

A flatworm's brain is about the simplest example. It can take inputs from sensory nerve cells and combine them into summed signals. Then, it sends these signals to the right outgoing nerve cells. This process coordinates movements from sensory input to muscle action. We see these very simple brains in modern tubular worms, leeches, and tardigrades. Animals use their brains to combine senses. They learn from what they sense. This helps them move and orient themselves. They can also make programmed

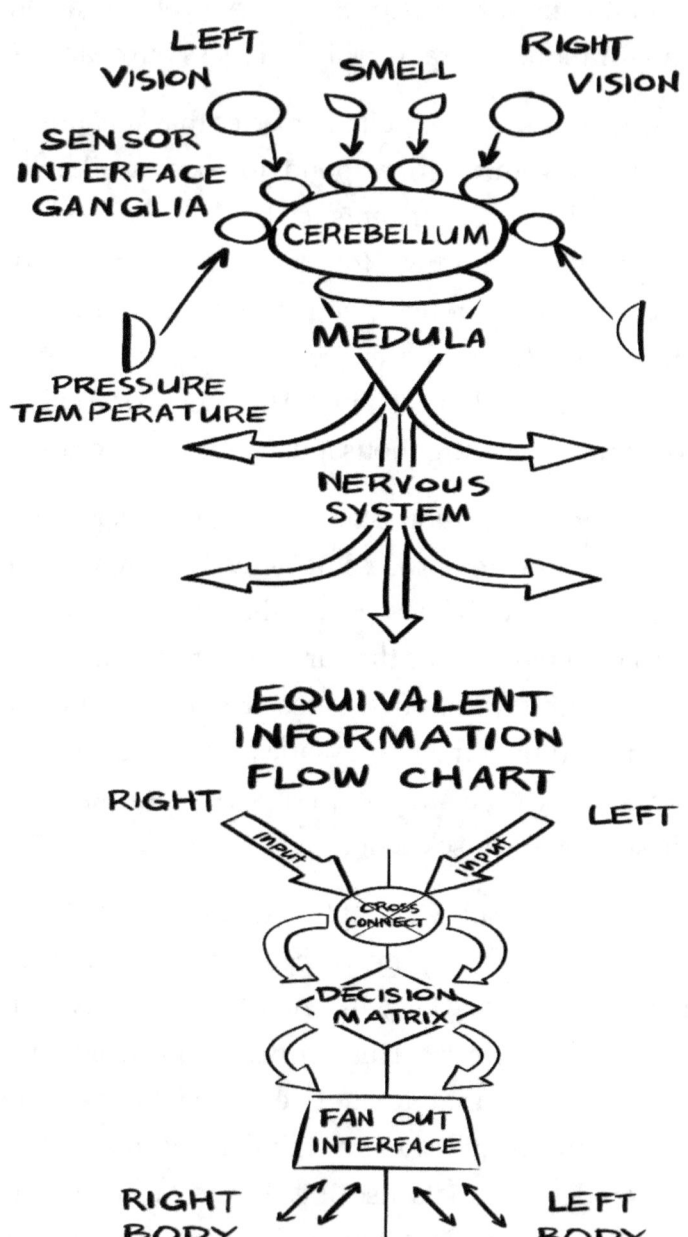

movements to stay safe and catch food. Simple brains transform sensory input into motor control. We can view the information flow as a *'feed-forward'* process. Here, information leads to action, relying solely on firmware to decide the outcome.

Sensors evolve to eyes, ears, skin (temperature and pressure), muscles (abuse pain), and chemical detection (taste and smell). Sensors are made up of specialized neurons with the appropriate chemical or physical functions built in by evolution. They send signals directly to special nuclei. This helps with preprocessing and connecting to relevant networks. Together, they interpret the information.

It has become symbolic information. This triggers resonant nuclei structures. These structures relate to memory and turn stimuli into action responses. Specialized nerve cells interpret these signals. They pass them to parallel circuits that send messages through the nervous system to all body parts. It maps one point in the body to one nucleus in the brain. This creates a direct link between the mind and body.

Coordination and control come from the brainstem, or medulla oblongata. This area connects with the spinal column and the rest of the nervous system. The brain becomes an entity unto itself and surrounds itself with special membranes. It takes a lot of translation and interfacing to send signals in and out of ongoing simulations.

This process is crucial for making decisions. The blood-brain barrier is strong. It stops blood from clogging the delicate nerve centers. However, it allows oxygen, glucose, and other important chemicals to pass through. Electrochemical actions work best in neutral electrical conditions. Small electric fields can help push large charged molecules to their rightful places.

The third transition marks a big leap in brain complexity. It adds neural clusters to create a more complex network. This network improves signal processing feedback. When the output of a process is folded back on the *feedforward* process, it is termed *recurrence*. This includes both positive and negative feedback. One increases output complexity, while the other reduces it. Most animals, after the first major life extinction at the end of the Ordovician period, 450 million years ago, gain this jump in processing power. Insects, which also appear after this extinction have recurrent brains.

Bees are brilliant. They quickly learn different types of art. They recognize abstract ideas and navigate to their goals. All this is possible thanks to their unique brains. We claim we are the smartest animals, and perhaps we are. But a bee can do things a human simply cannot.

Human intelligence demands an extended childhood, in which they can't even walk for a year. A bee is fully functional from the moment its wings dry and it emerges from its incubating cell. A bee can learn to navigate for

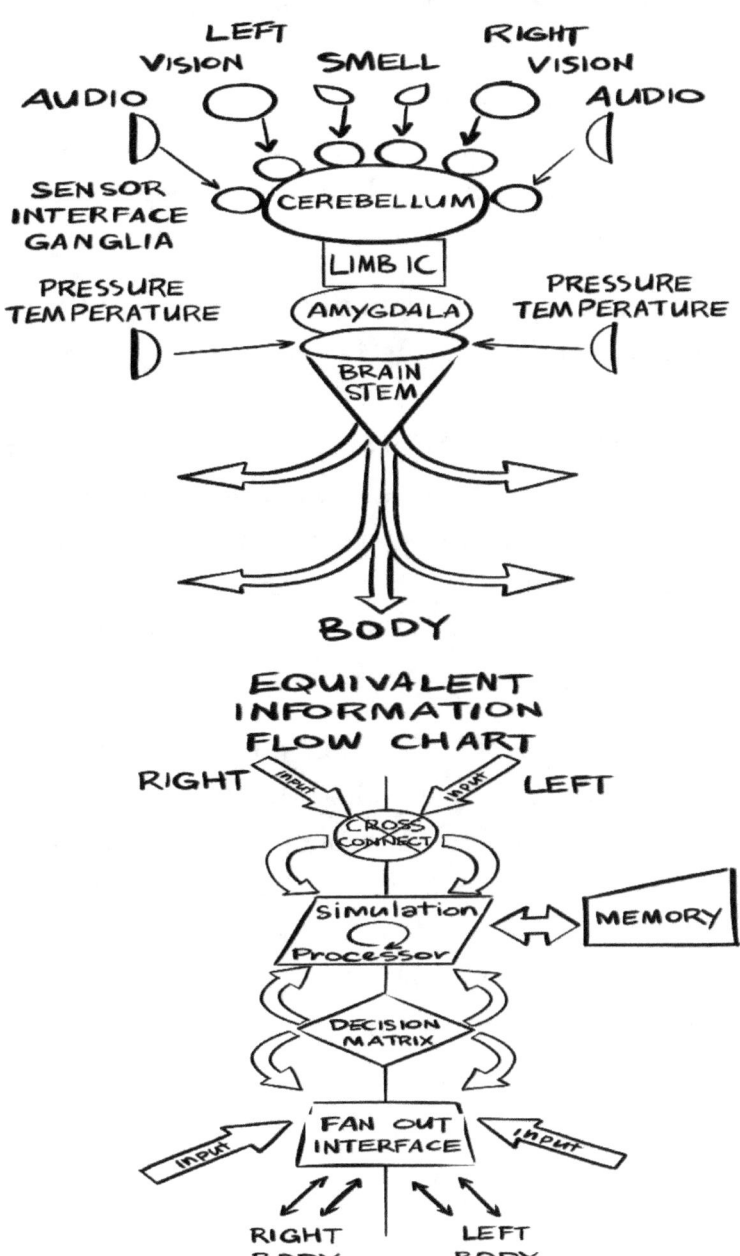

kilometers around its hive with less than 20 minutes of flight time. Its brain structure is preprogrammed to learn. A template guides where and how nerves grow and connect. This makes it ready to solve these problems right out of the hexagonal birth chamber. Humans still get lost walking to the bathroom at night at any age.

A jellyfish or a worm might not be Einstein, but they can tolerate a level of damage that would kill or paralyze a mammal. They can even regenerate missing parts, regrowing them with the help of neurons. Different animals have found success with various types of brains. Each brain type helps them thrive in their specific environment. Different environments shape various paths of evolution. Changing conditions also lead to distinct types of successful animal brains.

The fourth big transition for more powerful brains happens when we add major overbuilds. This creates many neuron nuclei that form dense, interconnected, multiple recurrent systems. Each system connects back to the others. This appears in animals surviving the Permian extinction some 250 million years ago. In this new version, information moves through the mirrored recurrent systems in cycles. It also has a recallable pattern memory.

This advantage helps the brain become aware. It learns to adapt and be flexible when facing new survival challenges. It can change its behavior when forced by circumstances requiring a new strategy for survival. Brains like this occur

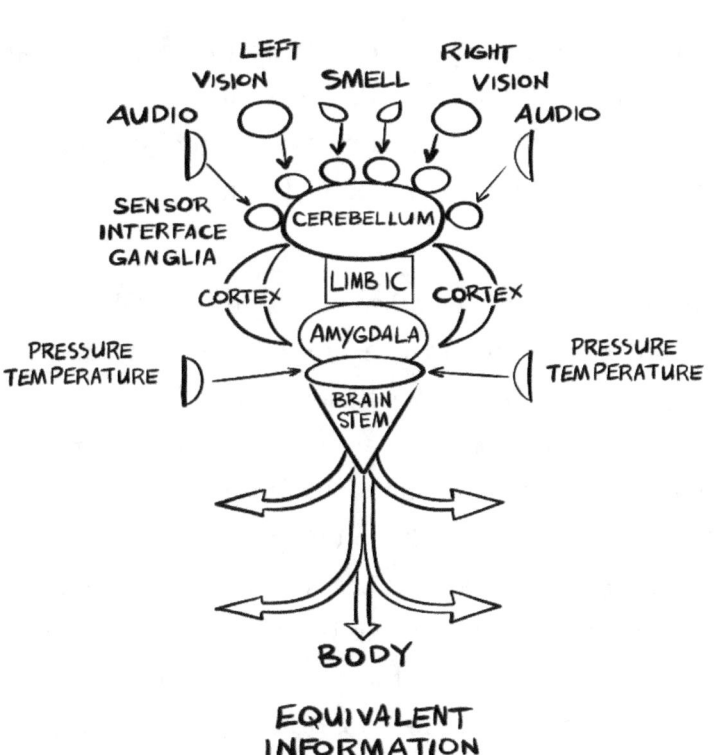

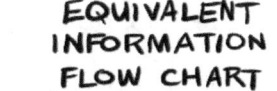

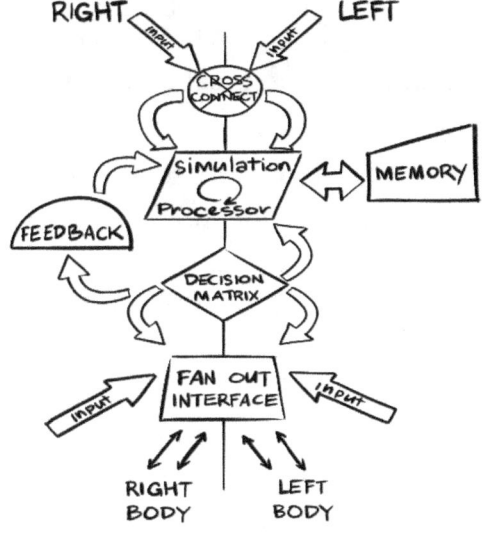

in reptiles and fish all the way up to birds and mammals. Dinosaurs, except for birds, likely lacked this ability. This may explain their extinction. When food supplies changed overnight, they needed flexible brains to adapt to their new habitat.

Multiple recurrences allow massive parallel processing of information. You can use the same information in many ways at once. Also, other closely linked clusters can identify relationships between different types of information. This helps animals learn quickly. It gives them a boost in intelligence. As a result, some species become smart about using and exploiting resources. They are able to find niches for survival that are not exploitable by lesser brains. Ravens and rats have changed their homes to thrive on something new: human waste. They are doing quite well, thank you.

These networks of repeating systems help birds learn complex songs and even some vocal sounds. This is why birds, rats, and dogs are great at learning what, where, and when things happen. It's why chimps can learn new ways to manipulate objects to solve problems and make rudimentary tools. Technology shows up in the evolutionary record as a natural step. It evolves alongside the brain, the eye, and the idle hand.

The fifth advancement in brains didn't add new parts. Instead, brain-blood membranes folded tighter, creating a denser neocortex. This means more nerves are getting fed and oxygenated than ever before. Even our closest relatives,

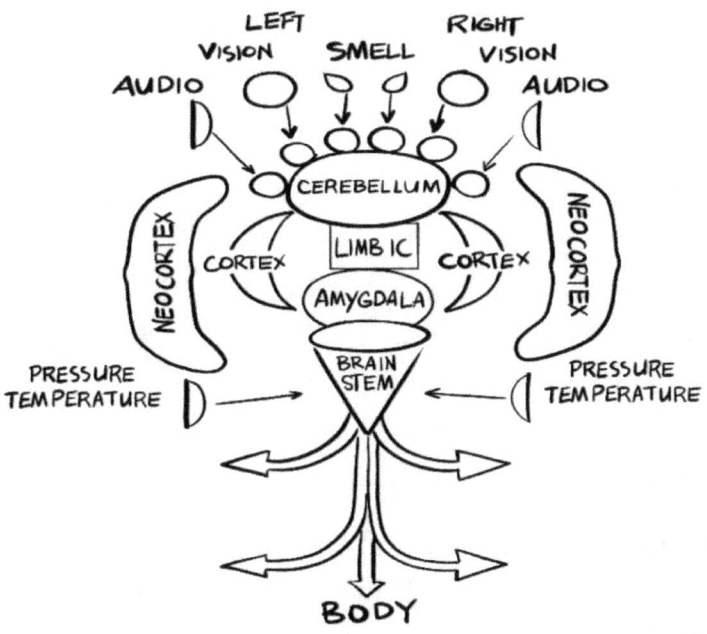

EQUIVALENT INFORMATION FLOW CHART

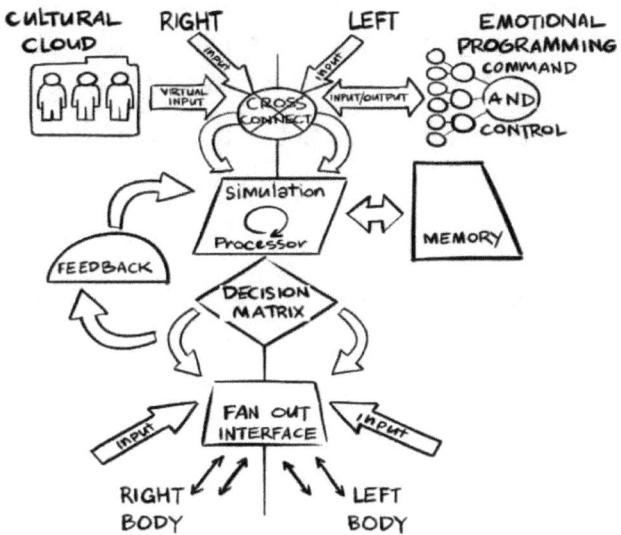

Neanderthals, don't have this structure. It's due in part to the versatility of a new kind of neuron making up an expanded and fortified frontal neocortex. This neural adaptation lets the brain change its structure based on its needs. It helps us feel good and reinforces successful survival.

In computer science, this is called reflection or what I call *'stored machine programming.'* The brain now relies on a complete set of software instructions. These instructions aren't physical and need to be programmed into the brain for it to function properly.

What's crucial is that relying on virtual programming creates highly variable behavior. This behavior can change in real time for various reasons, not just for survival. It may have started with that emotion but can shift easily. Hominids can program their own minds on the fly, but they can't explain it, only show it. We call it making up our minds or changing our minds. It's the ugly process humans go through to make decisions based on self-acquired information that makes us feel better when we get it right.

The key change at the fifth level is the need for a virtual link to a trusted outside information source. This means there is no longer a need to preprogram the fetus with just firmware. A human's behavior is shaped by the brain's structure. This structure is inherited and sets up how we act. The child lacks specific programming details. They don't know how the decision matrix is set up or how it works. They are also unclear about the algorithm it uses and how it interprets input and output.

A new virtual sensory interface supplies all this, similar to eyes, ears, and noses. It's a network that sends cultural programming info straight to the cerebellum as symbolic information. Virtual sensory survival information is conveyed by this new level of thinking. This is the quantum jump that makes humans stand out from all other hominins before us.

We have conscience and emotions. We can make abstract symbols that communicate ideas. We share and receive important information instantly through strong virtual links. I like to think of culture as the rumor mill bubble we live in where we are overly influenced by the people we listen to and gossip with. It's a virtual connection link directly between human emotional operating systems.

These five transitions help us make sense of our place among the stunning diversity of animal thinking and how we got our brains. Nature shows five distinct brain types that thrive in survival. One is not inherently better than another; each simply serves different unique types of challenges. However, it may be true that the fifth brain can use its unique power to eliminate all other life on Earth, or change all life to whatever is desirable. We kind of won the brain dominance game but haven't decided what to do with the winnings.

Each of these transitions is a set of evolved changes in the structure of information flow through the neural-networked brains. Each transition changes what the brain and nervous system can do. It also opens up new possibilities for adaptive behavior. The transitions build on each other. You need a central processing structure to enable repeated single-pass actions. Consciousness consists of repeated simulations without interruption under conscious control.

But the story of brains is not a ladder with Homo sapiens at the top. It's more like a fan where each branch can define a ladder of its own. Mammals such as whales and elephants have big brains. These brains help them manage their large bodies. They live in tight groups and care for each other over long distances and in a vast ocean.

Insects have small, specialized brains. Their short, active lives require structured activities to succeed. By their sheer numbers today, I'd say they have to have something going for them. I've seen mosquitoes that plot like scheming Wall Street bankers as they sneak up on me for a little midnight withdrawal.

The reflective nature of human brains lets us change the way we process information to suit each individual task at hand. A reflective brain can find the best way to handle new or changed tasks. It can quickly adjust how it processes information to finish tasks faster and more efficiently.

The human brain can reflect. This ability sparks our imagination, shapes our thoughts, and enriches our minds. It's often seen as spiritual. It also allows us to use symbolic language. This expansion helps us communicate and coordinate more effectively and efficiently. As our need for information that becomes more important grows, our processing bandwidth also increases. We have a natural urge to connect with and be part of a larger group, no matter its survival strategy.

But this mutation only frees humans from natural selection for the physically fit. The individual does not have to be strong, just the bonds holding the group together. Humans must fend for themselves not individually but in groups. They have no physical advantages, just a sense of vulnerability, a strong sex drive, and some naughty hands. They fix the gap in survival strategies with new technology, better communications, and smarter planning. We gather to make our plans. We sharpen some rocks and head to the canyon. There, we find the mammoths migrating. At dusk, when they are most vulnerable, we strike.

When our ancestors, the hominins, climbed out of the trees, it freed up their two hands for doing things other than hanging on to a limb. The hand and the human brain develop together. They become extensions of our ability to share information with those around us. Hominins begin to coordinate their vocal sounds with pointing hands and body gestures. This helps them share information more quickly and in greater detail than just by watching and copying.

Hands, face, eyes, and sounds generate a lot of information. This is especially true right after birth when the baby is at maximum learning potential. The human brain constitutes a dramatic jump in the amount of information that brains can process.

The neocortex in hominins evolves to predict a future that seems uncertain but is crucial for survival and success. New long axon nerve cells in the frontal neocortex add cross-connectivity. This helps connect to other brain areas that handle learned tasks. Like all neocortex's in hominins, the central brain models reality. It uses current observations to create simulations. Then, it projects future scenarios based on past experiences and the emotions linked to that knowledge. It's all based on the amount and value of information being processed and its associated emotions setting interpretations and priorities.

These new long-distance command and control nerves add organization to the reflective brain. They let specialty processing regions connect and activate across the whole brain. They skip important thinking pathways. Instead, they use options controlled by a conscious action, much like programming a digital computer with a live interface.

This new overlay neural network creates emotional consciousness, awareness, and intelligence. At birth, these abilities are not active. This is due to the absence of a program that connects the existing neurons into a functional network. All hominins have this structure except for one thing.

A change in hominin brains creates one that allows virtual communication. This connects to a different neural matrix, which controls the organic brain at a new level of complexity. When neocortex nerves become dense, they need more blood flow. At a tipping point, the brain lights up. This causes intense awareness and a tendency to alter its own programming. It shows the influence of the newly expanded neocortex. It serves as a matrix that links various specialized processing layers. This creates a virtual processor that is aware of time and can plan for the future. Additionally, it has a strong cultural connection for support and a smart way to manage randomness.

Fossil evidence for emerging humans reveals a rise in neocortex folding. This suggests greater nuclei density. Recent studies comparing genomes confirm this finding. For the first time, an organic brain evolves what can truly be called intelligent. Organic Intelligence (OI) is a computing machine's ability to create, change, and use its own complex programming. It connects to a larger information network and has the capacity to manage its flow of data for predicting future survival.

Humans now rely on shared abstract information. They exist in a cultural group that helps them learn and grow.

This group provides the support needed for both initial and ongoing programming. This new blank neural network brain learns from its parents and their culture. It connects through an abstract interface with unique symbolic signals. Now, with emotions involved, the information gets sorted. It becomes priority based on what the cultural program emotionally dictates.

Our decision-making brain follows an external algorithm. This algorithm aims to help the cultural group survive, not just the individual. Emotions act like a programming language. They shape our thoughts and help reprogram the neocortex. This change makes us less self-centered. When people need to change their internal wiring for a reason, it often starts with an emotional jolt. This jolt clears out old circuits allowing new ones to form, changing how the brain fundamentally thinks. You can change human intelligence by using the right emotions to guide the transfer of new information.

With OI, tool making and technology grow rapidly. Soon, humans can rely on these advances for complete survival. They use technology for strength and power, easily defeating top predators. They hunt down and eat damned near anything edible and are awful good at it. Our simulating mind can now solve the *'What if...'* conjecture, stimulating ingenuity and creative thought. We make things and then we make things better. We stop chasing food and start growing it in one place. That starts cities, economies, and psychosis.

As the top dog, we also need to be part of the necessary virtual information network. Our growing intelligence needs this for proper functioning. We form large tribes or communities that grow into cultures. This gives us strength in large numbers, more than enough to outnumber tiny Neanderthal family groups. This gives us a cloud of virtual information. It's made from past cultural memories that people can share.

They don't have to learn everything from the beginning, unlike other hominins. Cultures gather survival knowledge and dispense it accordingly. With teamwork and adaptable technology, modern humans can easily tackle challenging environments. Examine the Inuit, who thrive in one of the harshest environments on Earth. They simply do this with a little song and dance and some damn good technology for staying warm and keeping well fed.

The brain now only battles chance, like random events. It also faces the risk of sudden life-threatening situations, where there's no time for typical brain simulations or past knowledge to help. It must quickly react, intuitively make decisions, and compute a path out of harm's way with no previous experience or guidance. It has to come up with a strategy to deal with the unpredictable because emotions tell it to. We must obey the first rule of organic life. We must survive no matter what.

Chapter 5: Baby Emotions Drive Civilizations.

In the last chapter, I briefly explain how the brain evolved. Over about half a billion years, it changed from simple neural cells in a basic network to become human, programmable, and intelligent. Life on Earth is animated and diverse. Tiny multicellular organisms dance in the waves, while countless animals roam our planet. Each one faces a key question: *"Should I stay, or should I go?"* Survival is always on their minds. Brains must make decisions with the risk of paying dearly for bad ones.

What we understand now is that humans are not playing the survival game by the past rules of nature. Before humans, brains had built-in pathways and nodes for clear, logical thinking. The same solution to the decision-making is applied endlessly until it varies, and in so doing, creates an even better programmed brain, or not. If better, it survives to make babies, and it catches on. It's sort of like trial and error, except the trial is survival and the error is extinction. Change toward better adaptation is slow but sure.

Life is driven by what we call the life force. This is really just a tendency of certain organic chemicals. When they have the right conditions and enough time, they create more complex and rare molecules. Organic molecules tend to be more fragile as they get bigger. So, it should reach a limit in size and complexity based on environmental factors. But complexity can also lead to new solutions. For example, we can grow a skin to shield fragile chemicals and their intricate processes inside a shell.

Another life force appears where complexity must have one more attribute before life can truly begin. It must survive in the environment by some form of perpetual repair or reproduction. Reproducing helps an entity stay in its unique physical and chemical form. It makes new copies of itself before time and nature break it down. Life is a constant fight between being and not being. On Earth, it mostly supports existence when carbon levels are stable.

Life here on Earth must find the means for survival, or it simply can't exist. For over two billion years, stable conditions on Earth allowed the first living cell to form. When the protein-based cell evolved, it grew quickly in various shapes and sizes. It feeds on sunlight, water, air, and complex organic chemicals, including other cells.

Some cells get eaten but not digested. Instead, they evolve into specialized cells. This helps create more complex chemical factories. All this happens within the safety of a strong cell wall. By coming together into complex, supportive multicellular structures, cells specialize and synthesize. This gives them a much better chance of survival.

In previous chapters, we explored how evolution shapes animals and plants. They adapt to fit into their environments, which drives their evolution in specific ways. Survival for all mammals relies on mutations. These changes create diversity and different strategies. Then, reality tests each strategy through survival for continued reproduction.

Mobile life contrasts with fixed life. This is due to specialized neurons that generate electric and magnetic forces around them. Molecular electrical forces help sense, coordinate, and trigger key life activities. They enable the multicellular entity to move. This translates into a better chance for the affected entity to find food and avoid becoming food.

Survival drives the development of more complex nervous systems. This boosts communication and coordination across the whole structure. They focus on food, threats, and sex—the good, the bad, and the ugly aspects of life's challenges. They provide the means to create effective solutions for any survival strategy. This strategy requires the body to always take action and aims to eliminate luck from the equation.

Brains growing in complexity along with their host's physical form seem to dominate the path of evolution. Evolution is driven by successful survival, which complex brains seem to be better at. The better the brain, the better it can predict the future, avoid trouble, and make it through another day.

If it had only grown in size, then elephants and whales would be at the apex of evolution and intelligence. Smaller life forms can survive well without growing larger. They rely on their complexity for survival instead of just size. But greater brain complexity, upright walking, and free hands create a new option. This alternative to simple evolution changes everything. In one word, it is information in the form of knowledge driving technology.

To early technology practitioners, it is simply one of many ways to help find and utilize food and protect themselves from the elements. Technology appears in the fossil record over 4 million years ago in the form of shaped stones for pounding, cutting, and grinding. It appears alongside the use of fire to cook meat. This cooking boosts protein intake, which is vital for the growth and function of larger, more complex brains.

In return, the bigger brain keeps predators away by using brains instead of brawn. They cleverly use barriers, locations, and fire. They harvest prey better by observing and thinking ahead. They predict the future, which helps them move around and find food or shelter. Most importantly, brains create tools to make planning and doing these tasks easier. Survival has become a game of information and how to use it in the form of knowledge for surviving nature's nasty surprises.

Around 200,000 years ago, our hominid underwent a major evolutionary change. This shift introduced a new genome and a more advanced neocortex. The long neurons of the neocortex used for monitoring and control, multiplied in number and penetrated older brain areas. This setup allows us to function like intelligent machines that can rewrite their programming. Humans no longer can rely on preexisting firmware or preprogramming like Neanderthals did. Now, we are equipped to tap into a new sensory channel of information. This shift makes rigid preprogramming unnecessary.

Delaying functional programming until after birth opens a new information channel. The neocortex uses a virtual data link that runs on emotions, much like how operating systems are downloaded in digital computers. The neocortex adds two new features that replace the fixed firmware in hominid brains for early programming. A virtual interface links the brain to a larger neural network and it supplies the initial programming during the extended learning phase. This external virtual neural net connects humans for advanced information processing. It dominates and controls the neural net in our young minds. Humans think with more than just the neurons in their heads.

The brain has the ability to program itself in real time. It uses an organic operating system that responds to humans' complex mix of newly acquired emotions. Emotions act like the machine language of our brains. They guide how we handle, process, and store information for later use. Emotions are key tags that guide our focus. They help us think and sort information. We use these tags to categorize and rank results. As learners, we gain intuition. Our brains develop this skill as they join the bigger cultural network.

This new, huge external neural network has thousands to millions of virtual human nodes. These nodes represent human brains from multiple generations. They analyze, remember, and use information as shared group knowledge.

The key to survival now lies with the social group. This network shares a common language with fixed emotions. It downloads the thinking algorithms and information that all members need to develop. It's common-sense knowledge. This new way of sharing abstract emotions and symbols connects more people. They share common identities and work together. They plan for long-term results. Nothing in nature has ever attempted such a wide organic organization before.

Humans are not born with bigger or faster brains as perhaps their cousins are. They instead have a brain that is just an empty template that must be filled before anything works. When it learns unique social information in its early years, it becomes a loyal group member. This loyalty prioritizes the group's survival and interests over the individual. Group survival takes priority over personal survival. This helps the larger group gather and share better knowledge for everyone's benefit. Like I said before, there is more of the bee in our being than the ape in our shape.

If the new brain is born without an operating system, it simply cannot function. This virtual network of brains is now in charge of programming new members. For this new scheme to work, the new member brain must know about the virtual network. It should decode the information and use it as knowledge to finish its programming. It changes how long neurons in the neocortex fire or resonate. It awakens an unexpected awareness, which I term Organic Intelligence, or OI.

The virtual network must provide the new baby with a basic operating system and a virtual interface. After that, the baby turns into a programmed organic computing machine. After it boots up, the machine starts learning. It focuses on human emotions first. This is the basic programming language for the human brain.

Mothers smile, tickle, make faces, make cooing noises, and even sing to the newborn, giving the baby its first taste of emotions signifying good information. It mimics and echoes, showing it has understood and incorporated the information. A complex system of rewards and punishments shares more information. Emotional responses show if the new brain accepts and processes this information well.

It's a carrot on a stick scenario. The stick represents facial recognition and the reading of emotions. The carrot stands for physical rewards like caressing and praise, which show emotional love. This triggers a rush of dopamine, leading to smiles and joy. Learning has begun.

This continues at every stage, from childhood to adulthood. It lasts even after someone is an active, contributing member of the group. Learning and sharing information turn into an emotional experience. This experience gets tied to rituals and performances. This helps strengthen the social emotions that shape our thoughts and fuel our desire to belong. We all connect through shared emotions. The performance of rituals binds us together. This creates a virtual *'mother-net'* above our brain's neural network. It provides essential care and maternal love with indoctrination. This influences our decisions in the virtual world.

Breaking away from a group and embracing new social norms is possible. Deep down, our brains share a basic need: a virtual connection for knowledge. This need allows us to learn and adapt in real time. Additionally, it brings feelings of pleasure and a sense of belonging. It's a physical addiction just like opium. When we process information together and feel part of a bigger group, we discover our human need for freedom. This freedom exists within the safety and support of that larger community. This is the beginning of the fatal delusion of all addictions.

After emotions are mapped like pseudo-vectors, they fall along different dimensions of a Hilbert Space. These include laughter and crying, anger and love, happy and sad, and euphoric and depression, just to name a few. Once this mapping is done, the brain can adopt multi-dimensional algorithms. These algorithms use emotions like computer commands. They create neural activity patterns in the neocortex. This process changes the brain's chemical structure and electrical connectivity. The brain has chemical signals that help create and interpret information. This ability allows for reflective thinking and simulating reality. Consequently, now it can make faster and more accurate simulations of the future.

Emotions connect an individual's programming to the virtual neural network group. They can confirm or change decisions until the group's needs are met. When this happens, the group sends back confirming reward emotions. As a result, the brain gives itself a dose of dopamine for a job well done. The same applies to sex. The group uses fake emotions to control sexual practices. They override the preprogrammed firmware. This is often unnecessary and unproductive, sometimes leading to psychosis.

The new unprogrammed member needs to focus on large information flows. It will need built-in loyalty, conformity, or an addictive trait to accomplish this. It creates rapt attention to this learning, similar to how a baby is drawn to its mother's milk or how a junkie craves drugs. Mothers become surrogates of the virtual mother-group social image they are trained under. This keeps the learner focused on essential group programming and values. These are key to the group's survival.

A mother's emotional reward completes the task of creating self-identity and awareness. Without a mother figure, some learn how to reward themselves with substitutes. All humans do this when they're weaned and go out of the nest on their own. They learn to live with an artificially created virtual emotion that once was real. It's the key to self-programming and ultimate self-realization. It's also a fatal flaw in the whole complicated shebang.

I identify this virtual connection as the human's *'3rd eye,'* a virtual neural network posing as a culture. Social animals have been around for a long time, but this has a fundamental difference. A mammalian social group usually has one strong breeding male. This male has several females that cooperate. There are also many younger siblings in the family group with weak ties.

A culture has many adult males who work together cooperatively. They hunt, gather, and reproduce in a tight group of socially related families. They also share a single spoken language, history, body traits and self-identity.

Cultures store and share high-bandwidth information through symbols, both sound and sight. Our symbolic interface to the cultural network handles information in a highly compressed and encoded form. From birth, children naturally understand and remember these relationships. They react with pleasure, acceptance, and ease, often without even thinking about it. It's called tradition and intuition.

Humans are not just individuals facing nature alone. We are cultural beings shaped by our programming and the information we require. This programming influences our thoughts, views of reality, and expectations for the future. Shared communication, accepted knowledge, and a need for conformity to the culture tie together a new virtual reality. This reality is more powerful and smarter than any single being could become alone.

The new cultural network has an answer to the old question of mobility. Humans stay together in numbers until they decide to go. It becomes a trivial question that the newly evolving super tribe deals with quite handily. When it comes time to move, they don't move hesitantly one at a time. They don't just randomly choose their path or stick to the same route.

Instead, they use a cultural future-predicting machine. This helps them find agreement from connected members, guiding them on when and where to go. They always stay ten steps ahead of their less sure competition or prey.

When post-birth programming wraps up, usually around puberty, humans still have their individuality. They exist within a cultural shell that we can accept or dismiss. Is culture the dominant thinking machine in the room, or are we? Clearly culture can only be expressed through its individuals.

Early humans had to create large cultural groups, but they lacked enough people to do it quickly. As a result, their progress was slow. They formed small bands, often with fewer than a hundred members, and struggled to define their identities. It took maybe a hundred thousand years before humans could form groups in the thousands. But when they do, they will likely dominate their surroundings, resources, and food. Larger groups can easily subvert any rivals they view as competition.

Creating large, organized tribes that cover vast areas takes a lot of generations. Only then do the benefits of a programmable mind in a virtual culture really shine. Fossil evidence is limited, but humans spent the first 150,000 years in northern and central Africa. From there, they spread out to occupy the entire continent and beyond. At this time, the Sahara Desert used to be a vast savannah filled with game. These conditions had been going on during a long ice age that started over 450,000 years ago.

The long ice age periods in the Paleozoic and Cenozoic eras creates a dynamic climate. It leads to changing sea levels and massive glaciers that weigh down continents. This, in turn, causes volcanic activity that results in cycles of warming and cooling in the atmosphere. Asia and Europe were hardly habitable, while Africa was the riviera of animal husbandry. Giant mammals, both living and extinct, fill every continent. They fight hard to survive, and they do so in huge numbers.

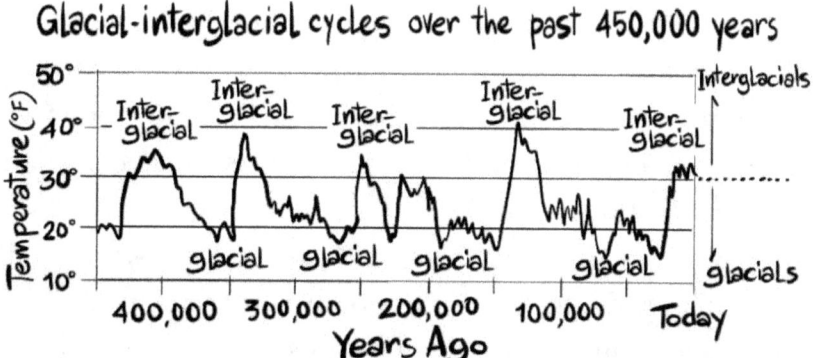

This meant the early humans were spread out in a natural garden of Eden where survival was not that hard. They had plentiful resources, so they didn't feel much stress to find alternatives. They also had to figure this whole virtual networked culture thing out, and that also took a while.

At first, there was no true language, but a strong need existed for symbolic information. At first, they were isolated and restricted to learning from observing the old society. Over time, enough new humans formed their own breeding groups. These groups became the first human virtual culture networks. As a consequence, the first advances in developing a truly symbolic language emerged.

At first, humans likely made sounds that copied animals and situations they observed in nature. They used their hands to gesture and act out symbolic scenes. This helps refine symbolic meanings. With an index finger to point, they combined sounds and symbols as representations. This created a simple language that everyone could easily learn.

I can picture how our ancient hominin ancestors felt about this strange new human baby. A new baby kept being born that seemed unfit for the tribe. They may have tolerated them if they contributed to the tribe as a whole. They probably were very fast imitators and could master old technology faster than the ordinary babies.

What likely bothered our hominin ancestors the most was how they interacted. They made strange sounds, danced like monkeys, and pointed at each other while continuing their odd noises and movements. This is not normal. There must have been some friction enticing the new human species to break away. This eventually caused a pinch-off effect from the old hominin groups. It created a new social species that crave their own kind and are plotting their own future.

They focus on their new virtual living entity. They shape its culture and create rules for how it interacts. They also build a complex symbolic language that ties everything together. Its histories, traditions, and myths are shared across the virtual network. This creates relationships that help members feel a sense of belonging. It also supports the sharing of important information as cultural knowledge.

Belonging is addictive and essential. It helps build a successful merit-based hierarchy with specialization. This strategy makes it strong and able to survive by not following the rules but by making them up as they go along. Cultures demand loyalty and sacrifice to the whole in order to guarantee the group's best chances for survival. This is upside down with all the rest of socialized nature and the source of our innate violence.

With this new way of life with improved survival, humans also gain intelligence and awareness. They have intelligence because they can change their own programming in real time. They are aware because they regularly simulate reality. They rely on cultural knowledge and virtual sensory input to be consistent. This way, they make long-range decisions that boost their survivability. This helps humans become more dominant and aware of the universe they inhabit.

They often have a natural arrogance that boosts their self-image, where most of their virtual communication is carried out. They dress up in clothes, masks, headdresses, and colorful feathers. They also wear animal skins and, oddly, paint made from mud and bright colored clays. They push these traits to absurd extremes, such as tatoos, to show cultural rank and power. This creates hierarchies which controls how information flows.

Human emotions control their programming. This is accomplished through ritual dancing, singing, and music. These are ways to express the rewards and punishments that help their brains learn and comply. Specialization goes so far that, at this point, some members actually do learn to dance or tell jokes for their meals. This is where culturally non-productive wizards, priests, and hucksters enter the picture.

They also quickly adopt technology into their shared memory and culture. They make advancements quickly and share them right away. This spreads new knowledge like wildfire. They can self-program by simulating emotions and confirming actions with rewards. This accelerates the use of technology that will soon make them the physical rulers of the world.

They already use sharpened stones, braided ropes, and woven baskets. They improve these skills quickly, unlike their hominin cousins. Technology fits into the new scheme of things even better than with earlier practitioners. With other hominins, tools had to be shaped by the slow and arduous process of trial and error. Once a new design is found to work better, they cannot simply convey this to the younger generation by teaching. It is instead a rather slow and awkward process of observation and imitation all over again for each succeeding generation.

Humans can program themselves using emotions. This helps young people learn quickly. Positive emotional support gives them a head start. They pick up complex social and technical skills needed to use and improve technology. No longer is hominin technology slowly developing for small family group survival. Now, with instant learning, large groups quickly gain an edge. New ideas spread in the virtual network like wildfire in dry grasslands.

Cultures that leave Africa due to population pressures or environmental stresses thrive in their new homes. They now move as a group, not just one leader with a few less affective members. They follow old game trails by tradition, but instead of hunting alone, they work together. They rely on technology like throwing rocks with a sling, using sharpened and hardened sticks, or snaring prey with twisted vines and nets. Humans invaded Asia and Europe with little competition. Their expansion was quick and complete, happening in just a couple of thousand years. Afterward, no other hominins survived.

Interestingly, the big migration of humans from southern Asia to Australia took about 10,000 to 20,000 years. They used low sea levels during cold ice ages to make the crossing. Once isolated in Australia, the Aborigines stopped evolving technically. They entered a survival stasis. Life became a moderate struggle and the environment stayed the same. This lack of change meant there was no need for improving their skills, leading to stasis.

They became adjusted and relatively satisfied with things, and as long as nothing upsets the apple cart, so to speak, progress never had a chance. Aboriginal people don't seek to control their environment through technology. Instead, they explore the spirit world of imagination. In this dream world, they find the necessary emotional rewards for living the good life.

If people want to return to a simpler life, they can do it easily. Just head to the outback, go for a walkabout, and see how long you can survive there. Aboriginal people have a deep spiritual life filled with dreams and virtual realities. They take the time and use substances to explore and enjoy this brain enriching life fully. They may not gain a condo by the sea for retirement, but when they die, they feel they have lived full, rich, and rewarding lives.

With no stress and no challenges, there is no drive for better technology, and no progress can happen. Humans migrating to northern Europe never had it so good. Struggle and technology are closely linked. The ice age brought big changes, not stillness. These extreme shifts lead to progress, which sparks cultural revolutions.

Humans live in communities made up of several families. Many adults in these groups carry advanced weapons, such as handheld knives, slings, atlatls, and bows and arrows. Most importantly, they have firesticks to make fire whenever they need it. They also have many special skills. For example, they know how to cook and preserve food.

They can knap stone tools to make fine axes and blades for cutting carcasses. They also tan leather to create clothes and useful tools.

They use all sorts of stone tools in support of an even greater technology: wood carving, basket weaving, and rope making. A mostly hidden history remains, with little evidence beyond fossilized human coprolites. These coprolites reveal that humans were smart in using and digesting local foods. Humans will damn near eat anything and will even make it into a delicacy in order to stomach it.

Their main technology focused on using available plant materials better. This included food, shelter, clothing, medicine, and making traps and nets. Humans team up to build communal traps that feed whole communities. They observe animal behaviors and use this information to their advantage. This strategy helps them spend less and less time and effort finding adequate food.

Humans are curious and observant. They learn by watching. This helps them find the best ways to get food. For example, they seek tubers and shellfish. These foods require skill to locate and harvest. Once discovered, technology like braided rope, baskets, and cargo slings helps carry the prize home. There, it is processed and preserved for the group's maximum use and benefit.

Everyone shares in the good fortune when any individual advances the group by producing more. Same for when they might produce less, as when healing injuries or with old age. The group knows where its strength lies and does what it must to maximize efforts with long-lasting results. They know how to survive. Now, their strong emotions help everyone work together toward success. Coordination through selfless belonging is paramount.

Future planning, teamwork, and sharing knowledge make survival a group effort. With many people involved and smart strategies, success is nearly guaranteed. We now share our brains with a network of other brains. This group forms a virtual entity that we often overlook. Still, we interact with it every day.

This new way of thinking about intelligent life comes from humans' programmable brains. These brains connect to a virtual world filled with information, symbols, and abstract ideas. Anything can be dreamed up as reality or truth if you simulate it in your brain.

The Buddhists and Hindus call this network of intelligence or connection to the cosmic intelligence, the third-eye into the mind. They believe this eye viewing the spirit world is fixed to their foreheads. They even mark it with a red dot to show it's there, somewhere. They have suspected this arrangement for thousands of years. However, they lacked scientific evidence to explain it clearly. They were close though, with the red dot actually sitting right in front of the brain's neocortex, which is the actual villain in all this.

Coincidence? Of course. Where else would you put a red dot?

For the brain's intelligence to develop, it needed to move beyond being an isolated decision-maker. Instead, it gained current and relevant information through shared decision-making. This is what we need to start the new self-programmable thinking process in the brain's decision-making system. It will help us create a vision for the far future. By syncing better with reality through a new way to share knowledge, the group makes better decisions. This leads to more effective predictions and actions that are tried, verified, and shared.

The firmware algorithm in the brain relies on what it inherits. It can only improve through repeated actions, which create alternate but related pathways in our neocortex. In new situations, it only uses the closest strategy from its limited experience. It repeats this over and over with minor variations. Progress is limited to those mutations that do it relatively right, by chance, and then pass them on by default.

Remember, DNA diversity happens when survival applies to a wider audience than what is needed. When the group survives, modified DNA shows up in different ways. This leads to physical and mental variations in how members perform and adapt. They earn their survival status, not directly, but through their connection to the culture.

Keeping the DNA fresh is a good thing. It allows for long-term expressions beyond just survival.

One of the early problems for any radical new species is that it appears in the midst of the older version and has to survive immediately on its own. Humans cannot do this as they lose the ability to be human without the culture training them. Without the members getting trained, how does the culture form in the first place? Some form of physical social splitting of the species must take place, but how?

I propose it is a pinch-off event. The first humans were part of the hominin social groups of their time. Training started to matter when one human gave birth to another full human. This results in a diffusion of humans within the hominin gene pool, which in itself is a long, slow, and painful process. Once babies are born with over half human genes, they quickly shift to the new model. Training then becomes formalized and integrated into the group's social structure. The band expands as families form a hierarchy of skills. After many generations, human DNA takes over. Modern humans emerge, cultured and skilled, ready to face the Pleistocene.

Blue eyes and blond hair were once seen as unusual traits. At that time, people believed they held a magical appeal. People could have just as easily suppressed and even destroyed them at birth as abominations. As long as variations align with emotional values taught by culture, they help improve culture. In practice, human behavior is often unpredictable. Decisions aren't always based on logic or reason. Instead, they stem from emotional desires and irrational actions shaped by culture. In a sense, we are like unpredictable quantum particles following rules of existence that seem contrary to nature.

Humans get bigger in size and live longer because those attributes have a chance of being expressed when food is good and wars are not. Group planning does this. In sexual attraction, cultural norms shape the next generation.

Sometimes, new ideas disrupt what was once accepted. These changes can impress their minority status on the larger group. They aren't seen negatively but as desirable, influencing shared beliefs. Suddenly, many unusual Nordic people appear worldwide. They claim superiority simply because they appeal to our basic sexual interests.

Social animals often experience a tipping effect when given new options. If these options offer a positive reward, they may shift away from their usual routine. Once about 20 to 30% of the group adopts the new routine, the rest quickly follow. This creates a consensus avalanche. This effect, where a small group can have a big impact, is well known in social animals, including humans.

Getting it right in our culture brings deep satisfaction. This feeling helps the neocortex remember connections for future use. Over time, these connections become dominant.

Memories are stored as unique patterns of resonating clusters. These clusters symbolize the reality we perceive through our senses. We picture thoughts using familiar images, including memories. These images hold complex information, like emotions. We recall and use them based on these feelings. This helps us create stronger patterns, making memory a dynamic process.

The conscious brain runs simulations of reality repeatedly. It changes the starting points or the algorithm until it finds a solution that feels good. This process gives the brain a sense of self-validation. We compare perceived reality with the simulation. This helps us correct and align it with our intuition. When predictions come true, it feels rewarding. This success strengthens the simulation and sets a new foundation for future simulations. We trust our intuition. This trust helps us make decisions faster and reach conclusions more easily. We don't develop conclusions; we jump to them.

Intuition combines all advanced thinking and future simulations. It leads to quick, rough predictions. Neural networks make good decisions even with incomplete or missing information. This approach reflects critical insights and shows intelligence and cultural connections.

Awareness means watching how we imagine different what-if scenarios. We also notice the feelings that come from each one. The most emotionally positive solution is then favored and applied as intuition.

Intuition and, more precisely, the programming of intuition, are paramount to the new human survival success. People can change their intuition whenever needed. They just need to reprogram the simple algorithm that links emotions to thoughts. Brain simulations can create an emotional reality. This helps us find intuitive answers. It allows for creative and unusual views on knowledge. These insights may improve survivability and reduce existential stress.

The simulation awareness process for predicting future events has some issues. This happens when the emotional programming fails. Symbology is key in storing and simulating reality. It uses low-bandwidth icons to carry high-bandwidth information. This information is tied to the specific emotions it represents. Information is encrypted and compressed. This limits recall information to small, representative amounts. However, emotions and virtual network connections can add deeper meaning.

Humans speed up evolution with technology. This technology is simply information and its resultant tools to create change. Then OI kicks in with planning futures communally. Programmed with cultural emotions, group compliance is key. It helps ensure group conformity and leads to better decision-making. This is vital for successful future planning. When the brain does this, it creates a large set of resonating nuclei patterns. These patterns are activated from a broad frequency of alternating current (AC) electrical signals. The brain oscillates at frequency tones generated by a natural feedback resonating process. The electrical patterns bear a strong resemblance to music or the arrangement of tones.

When two pendulums are hung side by side on a supporting string, they start to move in sync. One pendulum pulls the other along. Soon, they enter a resonating mode. They share energy, passing it back and forth like a football. One pendulum swings fully while the other stays still. After some partial swings from both, the first pendulum stops, and the second one swings fully.

Many conditions can cause this resonant behavior. The most common type is an incoming signal that matches the frequency of some chemical reaction. This reaction stores and amplifies the frequency in a feedback loop. This makes axons move and share resonance with other selected nuclei. Eventually, the brain's electrical activity sounds like the bells of Notre-Dame. Large swings in electrical activity get filtered, amplified, and selected. They result in some action or a new state of awareness.

A lot of electrochemical reactions follow a hysteresis curve. When they change with time, they follow one curve, but when they reverse, they follow a different curve. The result is a logical switching action. A good example is when you hear a funny joke. Your brain hiccups by following a hysteresis activity curve uncontrollably. This convulsion comes with barking sounds and muscle twitches. It creates a rapidly repeating pattern of resonances, causing emotional joy, sending bursts of dopamine shooting throughout the brain.

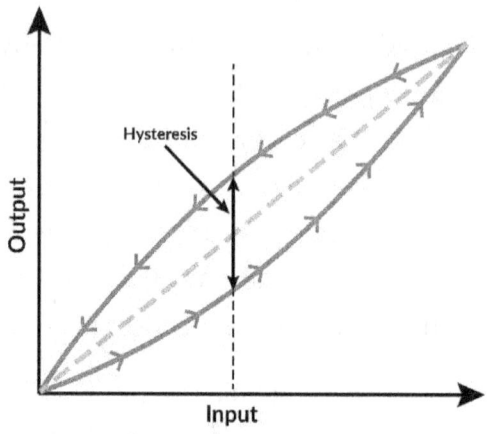

Resonances have a habit of getting out of control, causing misfires and possible seizures in large doses. Extreme laughing and crying are forms of intense feedback that can take over the brain.

They often lead to a reset or a clearing of thoughts, similar to how a calculator clears its registers. In severe cases, such as grand mal seizures or self-induced euphoria, the neocortex may not function properly. This occurs due to uncontrolled resonances that overload its circuits.

Connected neural nuclei create many electrical patterns. They are key to the brain's higher thinking. This happens when they connect dynamically across the brain. The neocortex's long dendritic neurons help with this process. This cross-connection works under emotional control. It boosts awareness and lets the IF-THEN logic happen. The brain acts like a natural computer. It uses chemical and electrical signals that ebb and flow. These signals help it learn and adapt through emotions. For the first time in organic history, it becomes smart and aware of its own thoughts.

A multistate, multi-cross-connected nuclei sends out strong signals when it is activated. These signals resonate with nearby connected nuclei. This creates a huge matrix decision-making simulation machine. It links feelings of good and bad, which then control our behavior. Patterns of resonating nuclei can create dynamic memories and learned actions. These often involve visuals and strong vocal elements like music, dance, and poetry. Basically, anything with repetition helps form complex thoughts.

These flexible patterns of electrical activity enable the brain to link sounds and images into equivalent thoughts. This forms a symbolic language that lets people communicate easily. It creates a rich cultural network where individual brains act as parts of a larger, collective mind. Brains display different levels of activity. Each level has its own purpose and response. They all combine to shape our personal reality. This is what justifies their high-calorie cost of operation.

There is now only one thing left for humans to worry about. Awareness and future predictions blend in our emotional minds. They highlight the true threat to survival. Humans become very aware that random events can happen outside their control. They are fascinated by this, calling it luck when future random events are fortuitously guessed. But without knowledge, they have to invent virtual information to complete the thinking task to the desired end.

I have illustrated this thought spectrum below. It shows how perceived threats to survival evolve from information into reality. It fills in what is needed for a smooth, continuous path.

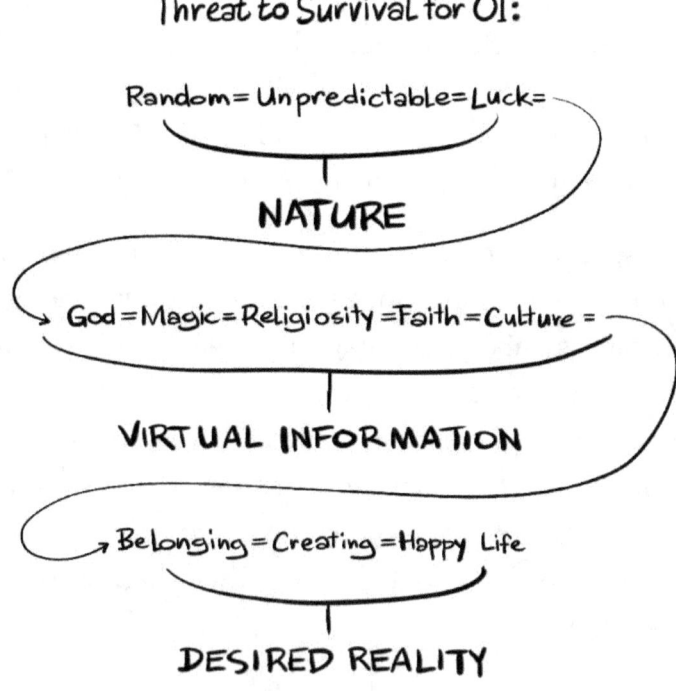

Notice the middle part side-steps into the spiritual world of gods and magic. This is new to natural thinking. It's one of our brain's quirks. This makes it our biggest weakness, but also our greatest strength.

Chapter 6: Technology, Hierarchy, Self-decorating, Entertainment, and Economics

In the last chapter, we learned that humans develop a unique neocortex. This starts OI by losing some old firmware and replacing it with something better. The human brain loses many instincts that help the mammalian brain make quick, life-saving choices as soon as they face danger. Humans are born with a clean slate with only a fear of heights and loud noises as leftover hardwired fears.

Humans need to program their brains with key information from outside sources. This prepares them to explore nature on their own. The cultural virtual network is the source. It holds the detailed knowledge that growing humans need to survive. Now, it contains much more technical information than ever before.

Intuition in each brain reflects the virtual brain's common knowledge or basic assumptions. The virtual cultural brain sets the limits to OI and defines what is considered true and productive knowledge. New knowledge can lead to new actions, but culture judges it first. If people do not agree, those actions are usually suppressed. Once consensus reaches a certain point, the entire culture shifts to the new knowledge base. This change redefines individual intuition.

If there is going to be any progress, cultural consensus becomes the new challenge for survival. New members with fresh ideas must first join the culture's existing knowledge base. Then, they can change the old culture's views on reality from within. Intuition comes from programming OI with accepted information from our culture. The networked culture is symbolic. It creates a virtual neural network that sits above the brain's natural one. This virtual network manages the brain's programming. It does this through conditioned learning and cultural enforcement.

Humans need cultural programming after birth to fully develop. This helps them join the necessary organizational intelligence network. This new intelligence is a virtual neural network. It connects brains using symbolic media mostly of sight and sound. Only humans can combine symbols and sounds to represent higher awareness.

This happens through associative processing. We use many specialized neural systems to process information. Then, another neural network, guided by emotions, sums the results filtering what's experienced. It increases processing bandwidth using emotional tags. These tags direct symbolic information between connected neurons. This interfaces the virtual cultural network with our internal network. Our brain interprets reality as culture sees fit.

The smart brain doesn't have to perfectly mimic reality to predict it. Instead, it makes ongoing guesses based on limited information about what will happen next. It uses what it thinks it knows and what it senses in the moment.

Intuition is the brain's best guess about the future. It combines information from various sources. The brain uses emotional programming to weigh this information. In survival, a near miss feels just as good as luck or being right. It's like dodging a lion's leap or finding a hive full of honey.

Our senses shape our conscious reality. This reality is both the source and limit of our intuition. The smart human brain makes fast decisions based on its first view of reality, no matter how much more information it could gather. It's the way a brain constructs a complex multiple-state solution, no matter the state of its various inputs. It generates a quick and dirty answer continuously that can be corrected later with better information. So, be careful not to trust your gut when considering things beyond your usual experience or awareness. If you're flying on intuition, be prepared to change course often in order to stay on a proper course.

From my experience, the brain can expand its insight to understand reality wherever it finds it. It can work at almost any physical scale and even grasp consistent yet non-physical realities. The only requirement is that it must be modeled in our brains in some self-consistent manner with the mental tools available. Our cultural intuition helps us understand reality at a physical level: inches, meters, miles, minutes, and years. First, though, our trained brain must spot the tiger in the brush before it sees us. We can picture this by shifting our view from the people involved to a mythical third person watching from above. The brain can change its point of view at will by simple manipulation of a three-dimensional simulation.

Everyday life works well with simple models. These models should match reality and make sense. As long as they reflect what happens, they are effective. We can expand this idea to other scales and perceptions, both huge and tiny. We just need the right modeling information. This can be shown symbolically, along with some basic organic brain simulation software.

For instance, students are warned when learning quantum mechanics that intuition is of little help. They are told to set that aside. Instead, they should carefully model reality at new scales and new realities. Using their imaginations helps them stay consistent and understand what our new knowledge or math shows us. Intuition is often limited by our past experiences. When we try to understand things beyond that, we run into problems. But it can be done, and humans do it regularly, sometimes without even being aware of it.

I personally struggled with quantum mechanics until one night I shifted my perspective. I used some mind-expanding techniques from the seventies to help me. I worked to simulate reality at tiny scales, like atoms, molecules, and fundamental particles. I was trying to prepare for my final exams in quantum mechanics.

I use a method created by Feynman. He was one of the smartest physicists in the last hundred years. He discovered quantum electrodynamics (QED) and won a Nobel Prize for it. He was a better physicist and bongo player than Einstein and Hawking combined. He also understood spooky forces at a distance very well.

In fundamental particle interactions, the math is enormously complicated. You can check if predictions are valid by looking at how interactions change with different coordinates, times, or other defining factors. When scientists apply relativity to equations of motion, it reveals that a single set of equations can describe various interactions in different vector spaces. These equations describe multi-dimensional vectors in artificial spaces, simplifying the analysis.

Beyond this, I will have to start writing equations. So, for now, let's take the high road. For the student trying to understand the basics, it turns out to be an *"aha"* moment. One dateless Friday night, I find myself absentmindedly playing with a metal pipe-cleaning tool. It has three arms attached to a single point of rotation. I imagine the tool as a symbolic vertex in a Feynman diagram. I can rotate the pipe tool in many different ways. Turns out there are only three unique translations. They all connect through rotations in a virtual quantum hyperspace.

In the diagram below, the upper-left one shows a simple proton-antiproton annihilation to a virtual photon. But for all the other interactions shown, all of them are exactly the same interaction, just rotated in hyperspace. Therefore, if you know how one works, you know them all because they transform so simply. In the world of quantum mechanics, this is a big deal that helps physicists solve complicated problems. We want to boil down complications to something simpler, something we can understand intuitively.

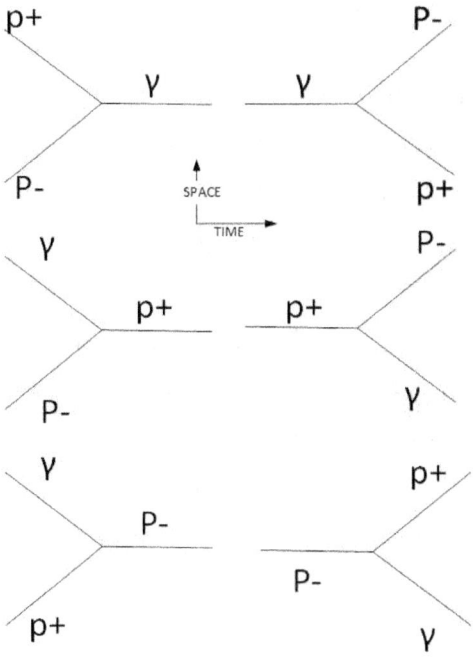

I was doing exactly that while visualizing the whole thing in a mind's eye simulation of a reality way beyond my actual senses. In fact, from this experience, quantum mechanics becomes second nature to me, or in other words, it becomes intuitive. When the mind explores a simulation, either in the brain or on a computer, it can move time forward or backward. This helps in testing different ideas and possibilities to find the best outcomes that match reality. Human intelligence excels at imagining what-if scenarios. It then checks these ideas against expected reality.

The brain adapts to different scales of intuition by changing its perspective. It adds rules and responses, creates a new modeling algorithm, and drives a useful simulation of reality. The more accurate and truthful the data supplied as information, the more accurate the virtual predictions. That's why we are so good at playing mind games like Go and chess. After a while, the brain gains a new intuition and continues playing by hunches and SWAG (Scientific Wild-Assed Guesses).

Amazingly, my mind didn't care if what I simulated was real or not. After all, it can be transposed from virtual to real and back again with little or no effort for a brain that does this for a living. It's all virtual and indistinguishable, sometimes blending fact and fiction in new and interesting ways. The key is to understand how our brain works like a computer program.

It learns from our culture. It takes in sensory information and creates simulations of the future based on cultural thinking templates. We see what we are supposed to see whether we actually see it or not because culture sees it. Our cultural virtual network interface controls our neocortex. It provides the main information link beyond our physical senses.

Human brains are fully programmable. They can adjust OI computing rules as needed. This is possible because our brains can change programs based on emotional system states. Commands in the OI programming language are just emotions ranked by reward and punishment. Even so, it's still intelligent and has the appropriate skills to prove it.

To understand this point, picture half-breed humans in a hominin tribe. They might show their value by being smarter and faster. For example, they could improve tools or create better slings for carrying babies. They seem to pick up on how things work much faster and make obvious improvements to things from the very start. They can also share this new information with another of their kind. This other individual hasn't seen the activity, but they still gain knowledge from it.

It must have really pissed off the leader of the hominin pack. He had to contend with a bunch of teenagers running around, convulsively barking at one another, acting wild and crazy at the same time. But damn if they don't come up with good ideas. One day, the little bastards team up against the pack leader. No one comes to help. He quickly loses his position due to their numbers and their surprising teamwork. Breeding rules shift to allow multiple males and females to bond monogamously. They coexist as larger groups in the same habitat. This setup quickly finishes the social pinch-off. It creates a safe space for new DNA to breed freely and securely, dominating the genome.

Once a human cultural group begins, they don't immediately adopt justice or equality. However, they come close, which explains their existence today. It's a natural requirement of any life-respecting culture and not of any strongman tribes. Cultural decisions should always be made on the merit of the question, not the results expected or the one most concerned with the outcome. Equality and justice is a natural part of OI culture. As is meritocracy which helps human cultures organize better. These qualities ensure the best survival conditions for everyone. Train with true knowledge, recognize best performers, and you'll never go wrong. Train with lies, pick the worst, and what do you expect will happen? Even Neolithic humans knew this much.

Technology helped hominins beat their competitors in strength and speed. This gave them an edge and ensured they could survive in any environment. It's in our DNA. No other animal matches our eye-hand coordination for fine motor tasks. This includes flaking stone, twisting twine, sewing clothes, weaving shelters, and using shoes. This was not a random mutation. Instead, it was careful selection over a hundred thousand years, or about six thousand generations.

During this time, humans developed their skills on the vast Sahara savanna. They learned to dominate and exploit their environment under near-ideal conditions. They used this time well. They worked on shaping the modern OI culture, focusing on its unique quirks and eccentricities.

The first sign of a virtual neural network of humans appeared early. This happened while working out the rules and procedures for male-female bonding. It was a clear necessity for maintaining harmony in a large cooperating group. They have to avoid constant competition for mating. It's an incredible waste of human resources with only one payoff, a child.

Ritual is the answer, as it is for a lot of the organizational problems faced by cultures. Rituals are key to compliant learning. They help share culture with all members with no debate. This builds cooperation and loyalty, which are essential for group survival.

Soon, the culture becomes full of rituals. Almost every part of daily life has a ritual that highlights its importance. These rituals connect deeply with personal feeling and emotions, shaping the awareness of OI. Only humans demonstrate these big brain activities. They make us laugh, cry, sing, and dance. It's all an act. Every night, around the fire, people make group sounds and movements. They captivate themselves with the basic emotions they create. They soak it up like a drug because it is one. Humans need to create rituals for their group awareness. This leads to competition, virtual rankings, privileges, and social hierarchies.

Rituals create a default meritocracy where the more competent dominate, not exclusively but in a leadership role. Cultures use formal rankings to impose order and predictability on their combined awareness. There comes into existence a virtual reality of natural ranking, almost like lions feeding at a kill.

People tend to form groups around a leader who is both likable and skilled. This leader encourages teamwork and makes decisions together with the group. However, when needed, they also take charge decisively. Oftentimes, they are appointed by the group they will lead, not unlike our modern selection of the team captain. In a true democracy, nobody seeks roles of power but instead are saddled with them as an obligation, like jury duty.

With rank comes two aspects, recognition and obligation. We enhance our creative simulating brain by linking it to cultural knowledge. This includes language, rituals, dance, and music. We also add body decorations, costumes, masks, and ritual stories. By the time we leave Africa looking for more territory as the one we had was drying up, we become a regular traveling roadshow. Musicians, magicians, and scary monsters tickle our fancy every night, driving everyone's imagination wild. Our hunting parties were the weirdest-looking and most fearsome by a mile. Stripping the land of its easiest resources makes it even more difficult for our resident cousins to survive using the old and true ways.

Humans quickly push out all competitors and settle wherever they can walk. Their technology quickly evolves during migrations to meet new challenges. They adapt fast and expend less effort in mastering new environments. They have high demands for awareness. They spend a lot of time and resources creating entertaining diversions. These include meaningful rituals and captivating stories about their culture's past and future.

This needs more specialized trinkets, tokens, materials, and oddities. These items grab attention and boost rank, respect, and value. Bright body paints help distinguish members of similar cultures. Grand festivals happen everywhere, and they are encouraged as part of cultural identity. Rituals and entertainment drive this celebration. Rare materials are traded far and wide. This helps blend cultural differences through shared goods and desires. This likely begins in Africa during the 200,000-year-old human incubation. For the last 50,000 years, they spread quickly around the world establishing trading networks as they move.

We know this from the anomalies appearing in the artifact records. Sea shells with drilled holes are found far from their origins. Some are hundreds or even thousands of miles away. Rare herbs, medicines, exotic feathers, seashells, and gemstones appear in odd places. You also find raw metals like copper, silver, and gold where they shouldn't be. What's going on so early in our history?

Humans, with their new survival strategy, start to create persons who don't join the usual hunting and gathering routines. Humans are so good at feeding themselves that surplus labor shows up almost immediately. Surplus labor is labor without a direct required purpose in feeding, housing, or securing survival for the culture. Surplus labor shows up as young members completing their learning phase but ending up without an immediate purpose. Hunting parties are already full. Gathering groups are already packed and plenty of food is being brought in for everybody, even the old, the sick, the lazy, and the restless.

Lazy drives technology. Restless drives economics. It's the human who decides to go out from their culture and contact other cultures, trading artifacts of no particular use other than decoration, that drives cultures. This invents the concept of money. Rare materials, like salt for food preservation and flint for fire making and obsidian for tools, become available as commodities at the right price. They also trade spouses, perhaps whole families, and above all, trade abstract information.

It's the invisible trade that likely has the biggest impact. When one culture finds a new resource, like herbal medicines, or learns about animal migrations, that knowledge spreads fast. Trading activities help share information as goods move between cultures. The overall cultural knowledge base increases for both sides of the trading wall, and everybody wins.

Using basic communication techniques, any two humans, no matter their languages, can always strike a deal. Intangibles like respect and hospitality can impact this process. However, both sides in every deal feel they've gained something. So, emotions remain positive and reinforcing. Rituals can boost trading results in many cultures. They help strengthen emotional connections so that trading becomes so popular that an offshoot of it takes hold even further. Some idle teen-bads, instead of trading, go raiding. They sneak around to find goods belonging to others and carry away anything they can.

This exercise shows that wants and needs have value based on special circumstances. It depends on getting two needed goods to the same place at the same time for a trade. Humans use virtual thinking to create a way to show value. This method isn't relative; it works for transferring value between places without moving the actual goods. Anything rare, easy to transport, and resilient to handling, usually works well if it has a known value.

All human cultures back then focused on body decoration and exotic drugs for fun and rituals. So, these became the first trans-cultural currencies. Beads, shells, colorful paint, feathers, dyed leather, animal teeth, and claws are all valuable items. Jewelry becomes the first money that is portable, easily secured by wearing, and universally desired. Drugs and herbal medicines are not far behind. Tools, weapons, and most assuredly mates also become trading goods. Bartering mates boosts trade connections through family ties. It also spreads successful DNA traits widely.

The result of all this is a virtual mesh neural network of cultures at a higher level of abstraction than the network of brains that make up each culture. This is a very thin and tenuous level that comes and goes as the well-known vagaries of economics rear their ugly heads for the first time. But this is the groundwork that must be done so that cultures do not end up butting heads with each other over resources. The culture that survives trades. It does not need conflict or the challenge of predicting uncertain futures with unpredictable people.

There is a deep-down positive emotion to live and let live among those of their own kind. But cultures put a lot of time and energy into ceremonies of rank. So, conflicts over virtual rankings can become brutal. Remember, the human mind is driven by emotional programming that controls our sense of reality. Duels to the death over ranking occur, and they disrupt the entire culture, sometimes wiping it out completely. People who navigate these cultural egos must be careful. They need to act as both traders and ambassadors. This way, they keep connections strong for the future.

Now, human cultures travel more. They stay connected through regular visits for trade and communication. This leads to a surge in technology. A fast way to share knowledge is now available. Nearby cultures share not just a prime location, but also languages and rituals. This brings them together into super cultures. They start gathering in very large groups. They construct permanent stone structures for shared ritual celebrations.

In the northern hemisphere, after fall hunts and food gathering, cultures gather. They meet at a common spot to trade and share before facing a low-productive winter. Then, they hunker down for the season. They turn this into traditions that strengthen the bonds within a larger culture. It also boosts survival chances significantly and is a key step toward social complexity. The sites are appealing year-round. They serve the needs of visitors who come for fun. They grow naturally into villages serving the local rituals, like the oracle at Delphi or the Kabbalah at Mecca.

To further this trading economic game, boats come into serious play. Boats are early inventions of humans so they can safely cross rivers in the Sahara savannah loaded with hippos and crocodiles. At first, they are just very big woven baskets. Then hollowed-out trees are made using stone axes and fire. Then comes the frame and skin boat. Better stone tools helped carve wood and make bigger boats. This pushed humans to get creative with technology. A clear problem emerged, and so did a solution.

Another key technology that comes into play, along with boats, is rope making. For any distance travel, it soon becomes obvious that the wind can blow your boat anywhere if sails can be controlled. Sails require ropes and cloth. Stone-carved wood, boring holes, hammering pegs, twisting fibers and weaving them into sails and lines all push technology to new heights. They used natural waterproof materials. Early boats are similar to the ones we have today. They sport a shape for gliding through water based on the shape of a fish and a structure of masts and ropes so a sail can be raised for free locomotion.

Humans coming out of Africa run right into a lot of waterways going off in all directions. The Mediterranean is now filled with sea water due to the last ice age melt. The opening of the Straits of Gibraltar and the Bosporus also helped. This created large, interconnected inland seas. The long coastline from South Africa to Southeast Asia invites sea trade. This trade is supported by the excellence of the Douwe. Today, the Douwe is still the most efficient way to transport goods over long distances. It can carry tons of cargo reliably for thousands of miles. All it costs is time and knowledge of wind and currents.

Humans view this situation and some large rivers as a guiding light. It encourages them to embrace boat technology and enhance their mobility. About 20,000 to 30,000 years ago, human cultures used advanced boats to explore the Pacific Ocean. They traveled north and south, reaching a new continent that had never been inhabited by any hominids. A similar expansion was happening in the

South Pacific archipelagos. This eventually led to humans populating every habitable island on Earth.

Mind, all these advances in human cultures take place long before the first city-states arise, kicking off civilization. I'm outlining the key milestones that humans reach after the pinch-off. These milestones set them on an irreversible path to OI and its impact, along with virtual cultural intelligence, or CI.

Modern humans develop temporary intuitions that work like instincts. Humans gain a flexible brain that can synthesize awareness. Awareness comes not only from senses but also from knowledge in the form of information. A virtual network provides information as knowledge that forms the culture. This network communicates through emotional symbology and equivalence. This network is solely responsible for early human brain programming.

The human brain is a programmable computing machine that possesses awareness of its own precarious future. It constantly simulates reality searching its fate. It predicts future events and makes life-altering choices. This is based on a larger and more detailed knowledge base than past hominins had. The culture shares information through visual symbols and repetitious sounds. This boosts our communication bandwidth. This also builds a culture rich in language, history, traditions, rituals, and shared feelings.

Human communication starts with non-verbal cues. It depends on reading faces, postures, gestures, sounds and body language. When we add symbols to gestures and speech, true intelligence emerges. Now, we can recall information from our shared cultural memory without needing direct experience. As humans gain the ability to control their cultures, it can be said that the cultures themselves become intelligent. They are complex networks of information links between nodes that process organic data. At some level of complexity, awareness is inevitable.

Cultures can develop personalities that process information and form opinions. They do this like individuals but are limited by the lack of diversity among their members. The more diverse the culture, the more imaginative and dynamic its solutions to the reality prediction game. The more similar, the less need for accurate predictions as the bet is pretty much already fixed by other issues outside of the game. If nobody in a car knows how to drive, it is pretty much headed for disaster no matter how much they agree about who should be driving it. If somebody knows best, they demonstrate that fact, and meritocracy takes over.

Paleolithic humans took important programming steps that show how their brains evolved. This evolution unerringly led them toward civilization and the upcoming singularity. This helps us understand how different OI thinking processes connect and influence each other. We need to understand our group thinking. This is key to reprogramming it for surviving the singularity and the next pinch-off event. We can predict what the human brain will keep and what it will discard. This helps us trim our cultural limits by making new assumptions.

Above all, the new human needs to be able to outthink and outwit AI when it comes down to who inherits the future and why. The why should be obvious, as it is the primary objective number one for all organic nature: survival. We know something is alive if it fights to maintain that life. This is as old as life itself, so it will probably stick around after the singularity.

Sex is another necessity for survival, but when modified by technology, it becomes entertainment, pure and simple. Sex unbounded making children is no longer tolerable. Procreation will be forever separated from sex after the singularity. No more unplanned babies and no more adoptions. Parents will need to get a license, and then there will be big restrictions on competency and commitment. So, we don't need to worry about sex being a highly controlled cultural obsession ever again. The only worry now is how much the *orgasmatron* will cost.

Males and females in the early days of humans are pretty much equal but separate. Partial apartheid of sexes at the time made a lot of sense and still does. We all have to deal with life and its challenges. So, it's best to do what we can to get along with it. Sex is like eating, drinking, breathing, and defecating. It's required by our bodies, so get used to it. We need sexual expression as part of our culture and our innate survival. Without it, we risk falling into biological confusion. We may struggle with strange feelings and wonder why we can't get enough. The chemical instructions for sex come from an ancient genome, so we can't alter it as easily as, say, eye color. But it will not be essential after the singularity. We already have more than enough humans sucking up resources and must limit their procreation. Technology will manage it better than the old way. It will extract the fun without paying the pregnancy price.

All current human children need to download information from culture through face-to-face learning. They must be indoctrinated into the specific culture by an inherited addiction of loyalty and conformity to a larger group. The human brain runs on emotions. These emotions use the brain's reward system to confirm and follow what brings rewards. This is a fundamental survival need for the culture, but not for the individual. All human cultures and subcultures need conformity and cohesion. This helps them survive together and enjoy the benefits of community. The self needs no such limits to survive.

The big benefit of this is the extra time and energy that a tech-savvy culture gives us. We can use this surplus for activities beyond just survival. The culture survives by properly indoctrinating new members. They spend their free time on activities that build trust and loyalty. One activity might be raiding economic resources from distant cultures for fun and profit. Another idea is to build public works projects. This could mean piling up rocks for a big celebration. The rocks might also be used to hold down the great leader's body while everyone celebrates.

Rituals from birth impress these ideas on the young. They bind the culture together and create a shared identity. This common belief influences their thoughts and activates their dopamine circuits. This builds blind loyalty, uniting members into a strong group. This ensures they will sacrifice for the greater culture. But it often needs constant reinforcement to stay valid. This requires familiar and comforting rituals. These rituals push group conformity even further.

This behavior is a survival tactic of cultural intelligence (CI). It helps CI thrive, even against other cultural forces. A specific culture isn't crucial for surviving the singularity. However, a different kind of culture will be needed to manage it and to support the pinch-off. Cultures that formed before and during civilization are often the most rigid. So, they are the ones that are likely to change or disappear completely. We need new, flexible cultures. These should focus on seeking truth instead of claiming they already have it. They can no longer insist they are the only ones who can understand the truth.

Most importantly, humans will still live physically together in cliques or clans after the singularity. Clans cluster together to make cultures. Trade and ceremonies connect cultures, helping them grow and advance technology. The more they share rare goods and valuable information, the stronger their cultural knowledge base becomes and the more advanced their technology. Cultural intelligence allows clans to be virtual, not needing physical presence unless so wished. Online groups centered on cultural memes often hold annual meetings. At these events, members wear elegant costumes and parade around ritual-like. This is a clear example of a newly emerging virtual culture.

Given OI awareness and the CI security, surviving now becomes an abstract battle with fate, death and probability. Humans track and use their internal simulated reality to gain an edge on the future and how it most likely will play out. Awareness helps them see that randomness as a natural part of life. It comes into play when they try to predict the future. They become fascinated by it. They invent games to guess random future events for fun and profit. They also spend a lot of idle time speculating about flipping a coin.

People mainly communicate through a virtual third eye. This means they use facial expressions, visible emotions, gestures, and symbolic vocal and visual language. This highly compressed communication at the clan level is by ritual and ceremony, using all the above means to connect its members. They form a mesh network designed to store and share survival knowledge. This cultural database serves as a creed, connecting people through shared beliefs.

The brain is a network made of special clusters. These clusters talk to each other using specialized neurons, a design shaped by evolution. Humans develop a second brain layer called the neocortex. This layer controls how clusters connect for creating resonant patterns. The neocortex sets up a mesh of mesh networks. This is extended by the third eye virtual communications interface to other nearby brains, setting up a clique or clan of brains. They all share in a higher virtual mesh network. These groups can grow and split, forming new clusters of networks. They communicate using shared information through abstract virtual symbols. Our brain can create equivalencies like, 🐕 = dog = 狗, for damn near anything, real or not. That is its power. Even when we think outside the box, what we discover is that there is no box, just equivalencies.

The upside to all this is humans are better at surviving by a long shot than any other animal on earth. No longer do they need to increase the survival database if survival is already as good as it can get. Technology now sets the limits to survival. It can be learned and shared quickly across cultures. This helps all humans adapt equally to any new survival challenge. But because it is now a virtual situation, it has no self-checks for truth or validity anymore. That's all been taken over by the database itself. It's called the arrogance of information. Just because it exists does not make it true.

Information itself, good or bad, true or not, provides its reason for existence by merely existing and nothing else. What's in the database feels true enough for the brain. It thinks it's vital for survival. This is especially true if the brain perceives a fake abstract threat as real.

And that is where the whole advantage of becoming an OI human suddenly disappears in a flash. Selfishness is a universal survival mechanism. All animals are naturally selfish to some extent, and so are humans. In a culture where image matters, people learn to deceive. They tap into others' fears, emotions, and needs. This helps them gain rank, prestige, and wealth. They often use flashy clothes, manners, and rituals to show off their status. All satisfying a simple animal survival strategy of eat first if you can. Everything fair and just falls apart when a human becomes a predator. They steal the culture's job of sharing vital survival information. Now, it's about the predator's selfish survival, disguised as a fake external threat to the culture.

An aware human fears what they don't know. They worry that the unknown might surprise them and strike when they least expect it. This fear often overwhelms them before they can even name it. People feel that if a culture provides knowledge for survival and safety in numbers, it should be trusted, no matter what. The word belief comes in here as a derivative emotion that trumps all other emotions. Extending trust and loyalty to the culture earns them the satisfaction of belonging and being accepted. This is paramount for hominins belonging to the clan. Fear of the dark is a negative emotion similar to punishment and can be used to program brains just like all the other emotions.

A cultural parasite first hijacks the cultural database. It fills it with false stories and myths. These tales claim to show more intelligence and truth than the old ones. This new knowledge goes against intuition. It tells believers to do specific rituals known only to the new priests. These rituals can help with luck and improve their odds of survival. The intelligent brain with manipulatable emotions becomes the gullible brain all set up for clever control or in its extreme, slavery. As slavery later becomes a violent feature of civilizations, I like to call them zombies instead. Zombies are the walking dead. They act by unnatural forces and blind fear. Their goal is to support the human parasite. They do this by turning uninfected brains into more zombies. This grows the horde, which feeds the parasite. The parasite becomes greedier, seeking more believers to control.

Lies can form belief cults. Predator practitioners claim to be saviors. They attach themselves like a parasite. They aim for smart people who trust their emotions. This is their main weakness. Trust is an absolute necessity for forming any group larger than the simple family. Human parasites infiltrate and feed off the clans' emotions. They take control of public discourse, acting like a virtual third-eye interface that connects only the true believers. They clearly aim to dodge the merit system, allowing them to rise in power without justification. This gives them control and profit for their exclusive gain. They even extend their lives at the combined expense of the clan they exploit.

Humans have a built-in flaw in their programmable brain. It is prone to being gullible for a reason. It's a weakness that brings people together, but it can also lead to denial, alteration, or control over vital cultural survival knowledge. Instead, simpler ideas take their place. People go along even knowing these ideas don't really affect the whole clan's survival. False knowledge only benefits those in power, often called priests.

We've traced life's journey from hydrocarbon molecules in a watery sea to humans. We've separated ourselves from our biological ancestors. Now, we represent the first organic computers that can self-program for simulating and solving any problem we face. We are about to encounter a new threat to our survival self-made by our dependence on technology. This needs a new solution mechanism never before experienced or anticipated. Let's look closely at who we are in this crisis. What are our strengths? What are our weaknesses? What actions must we take? What needs to happen? By answering these questions, we might find a way for our species to continue with earth's billion-year destiny.

For certain, belief cults will have no benefit for surviving the coming singularity. The singularity is pure technical, and it will take a science-based approach to survive it. Reality is about to rear its ugly head, and we had better be soundly footed in raw ability in order to face it on its own terms. The solution to an existential threat from technology, which is the one and only reason we are even here, is simple. More and better human use of information technology leading to the full convergence of our new cultural networks and science.

Humans pretty much had it made about 10,000 years ago. The continent-wide glaciers were receding, and it was getting warmer. The seas were also warming, and fish and wildlife were proliferating chaotically. Humans lived in small villages located in great spots. They ate well and had plenty of free time. Their lifestyles were healthy, and they belonged to supportive groups. Many lived long lives, enjoying them to the end. They traded goods over long distances and seemed content and peaceful, with no signs of major conflicts. Those would come later when the stakes got much higher.

They invent symbolic value for desired goods and use it to further trading, creating a simple barter system of commerce. They create the concepts of private and public property. They set universal rules for conduct. They also provide justice through group awareness, enforced by consensus.

They invent separation and specialization of labor, increasing productivity by thousands-fold. They establish unwritten social rules of public conduct as part of their common knowledge. They create the school teaching idea.

It uses experts to train the next generation. These teachers share cultural knowledge and tech skills suitable for their age.

Both sexes have the freedom to do as they please. There is no strict separation, and evidence shows little bigotry or misogyny. Males usually learn to fish and care for cattle. They fight off predators and do heavy work. In contrast, females often process food, weave baskets, and care for children. They also sew and handle tasks that need fine motor skills and focus.

Women fought as full-blown warriors alongside men, and men dug roots and tended children alongside women. Sexual freedom was abundant with no prohibitions against willing adult behavior. Stress and depression were almost unknown.

One of the most important technologies they perfected was, surprisingly, genetics. They started domesticating animals around 30 to 40 thousand years ago. This was when they began settling into small communities and adopting specialized lifestyles. With that much time to work it out, they created whole new species of animals and plants that never existed in nature before.

I chuckle at the so-called Genetically Modified Organism (GMO) fear among the politically correct. Some think altering any genetics is a bad mistake. But really, it has been the key to successfully feeding people for the last 30,000 years. You're a little late, thousands of years late, to complain about man-made organic plants and animals. Just look at the chicken or the milk cow, neither of which ever evolved in the wild.

A key benefit of this technology is the knowledge gained about genetics for humans. It can be used perhaps for breeding specific human traits, such as selfish thinking. They were the first to realize that male traits come from the male side, while female traits remain with the female. This will matter later as male dominated civilizations arise. They need to track heredity carefully to validate inheritance. Choosing mates and determining offspring gender, birth order, and provenance can become very contentious.

This went so far as to include the inheritance of physical and mental characteristics as well. Some female traits, like skin, eyes and hair color, were preferred for their seeming sexual appeal. This didn't help with reproduction or survival. Instead, it only maximized value in exporting wives.

Prostitution, as we know it today, began much later with civilizations, not Neolithic cultures. Neolithic cultures had a unique view of sex. They considered it a therapeutic activity and a shared enjoyment. It was easily accessible when needed and was never used for profit. That would be considered an insult, like charging value for health information.

Prostitution and paid health care show a harsh side of how cultures can go wrong. They exploit a loophole in our awareness, letting anything with valuable be turned into a product. If sex has value, then certainly life itself is vulnerable to a profit motive. Hence, slavery is just efficiency taken to extreme by civilization.

But during Neolithic times, another great thing happened that would equally change the world. They created, recited, and pantomimed stories to symbolically share cultural knowledge. This also offers pure entertainment. It generates emotions for their own sake. Plus, it encourages group participation and bonds culture. These are key parts of the new human need for survival in larger groups. We rely on our cultural network, where brains connect in loyalty and cooperation. This happens at a virtual reality level, guided by emotional feedback.

Clay is fired for the first time. This creates figurines of favorite heroes and characters. It also makes pots that are vermin-proof and waterproof for storing food. Furnaces are being made, proving humans don't just make fire; they can make it dance. It almost seems magical what high temperatures can do to certain clays and minerals. Mining of ores begins in earnest to find the materials to feed furnaces. They invent the bellows and figure out how to create high-temperature burning fuels such as charcoal.

They advance technology rapidly by replacing stone tools with metal. They begin mining and smelting many different metals. They start high-temperature firing of pottery and, with even higher temperatures, make glass. They make much finer sewing needles and stronger thread from cultivated fibers. They make cloth and rugs using weaving looms. They sew together tailored and layered clothing for cold climates.

They enhance leather tanning chemistry, creating a wide range of in-demand products. They improve fish hooks, rope, nets, and boats while inventing commercial fishing. They invent the plow and, with the ox's help, improve food production by a hundredfold. They tame animals, especially the horse, bumping the available power at hand from one manpower to one horsepower, a factor of about 10. They invent horticulture and breed fruit trees and other modern foods like corn, potatoes, squash, wheat, barley, and oats.

So, what happened, you might ask? Because I haven't seen anybody living like that today. There are solid Stone Age cultures found within the last couple of hundred years that could not fire pottery or smelt metals at all. They are remnants from before the Neolithic era. They survived because they were in isolated, hidden places, away from mainstream cultural forces. If there is no need to get better, then why go to all that effort?

The answer is that the culture over the hill is doing it. And the race for money, metals, energy, and power has just begun for humans. They'll call it the race to civilization.

Chapter 7: Civilizations, Religion, Technology, and War

Humans create culture to survive better with others. This virtual connection helps them learn and organize. It allows them to plan and act together. Working in harmony increases their chances of survival by sharing knowledge and focusing actions. Humans are born without instincts. They must learn to communicate and connect with others first. Then, they use it to gather and use information to survive.

Humans are uniquely shaped by DNA to form large groups. They use emotional bonds with these groups. They share important information with new members. They connect through rituals, music, dance, and vocal entertainment. They mainly communicate directly, face-to-face with expressions and vocal stress. They use emotions and symbols to share key information.

The culture provides new members with information that is colored by the interests of the culture. The cultural virtual network becomes a meshed neural network. An intelligent virtual networked brain rules above individual organic ones. This brain has an unlimited influence over all its nodes. It forms a new organic awareness and cultural intelligence, or CI.

CI comes from the way individual brains connect in a mesh network. They share a common set of identifying information. Emotions are instilled during early nurturing. Later, they help spread cultural information, leading to automatic acceptance. It is the programming language for human OI. It also drives the culture's survival. Group emotions come into play, and group dynamics affect cultural programming. This shapes the overall cultural database that defines cultural awareness.

Virtual databases of cultural knowledge should be true and accurate. This accuracy is key for gaining real benefits. It also drives successful human cooperation, which is essential for progress. It keeps the group together as long as it is perceived to be worthwhile. But only if the information is proven to be good and true will it have any tangible effect on reality. As a result, religion splits from natural thinking.

Social hierarchy is not based on who is the fittest. Instead, the virtual culture shapes it. Strength for the group shifts from bluster and muscle to brain and cunning. Power shifts from dominant bully to smarter huckster. People can become prey to others who usurp their culture.

Some people like to chat their way into food lines and onto the stage. They choose this over making helpful contributions. This is where an unexpected negative side effect of human culture comes into play. The idea of many cultures living in peace and harmony is appealing. They can trade, learn, and grow together. However, this vision has a deep flaw that has always been there. This weakness prevents it from ever becoming a reality.

Virtual connections between people create a network of intelligence. This network relies on sharing accurate information. When we share honestly, our overall intelligence grows, which boosts our chances of survival. People find it hard to distinguish between true and false information. There's no instinctive way to do this.

We can assume that one well-known characteristic of hominins going back to the great apes is the surprise attack by deceit. Hominins, especially young ones, often play tricks on each other. It's like they're practical jokers. This behavior helps them train and bond socially as they grow.

It also puts human culture at risk of a deadly disease. This happens when people intentionally fake cultural information. Doing so can lower individual survival chances and hurt group success. The goal for these fakers is to dominate the meritocracy hierarchy quickly before it is challenged. It seeks personal gain through deception, greed, social status, and sexual advantage. This happens without considering the detriment it creates for others in the culture.

In the last few chapters, we explored how humans thrived for the first 200,000 years of their existence. This brings us to the point in human history where the benefits of cultural survival were really starting to pay off. There was a huge population boom with a lowering of infant mortality. Whole new technologies were being developed. Small villages practicing specialized horticulture were appearing everywhere. They modified nature to meet human needs. This helped provide a stable diet for babies, making them grow bigger and healthier.

Human progress brings more free time. People use this time to share information and travel to celebration centers. These places host large cultural and cross-cultural groups. Here, they trade, share rituals, and exchange all kinds of information, whether true or false. People track the earthly calendar by observing the stars. This tracking becomes the focus of many annual celebrations and the design of their structures. The future is more predictable with good information, and the steady heavens are marking time very accurately.

As I mentioned before, humans are not only aware of what they do know, but also of what they painfully don't know. The main motivation for all is survival. They want accurate data to predict the future. This way, our simulating OI brain linked to the virtual CI brain feels safe and happy avoiding future surprises. Awareness from the newly connected neocortex means real experiences often don't match their predictions. They encounter many more things they don't understand than things they do.

They watch nature closely, especially other predators. They learn what drives these animals and why they act as they do. Humans quickly used this knowledge. They used it in rituals and ceremonies. This celebrated their cleverness and showed they were better than animals. These rituals are just entertainment. They involve dressing in elaborate costumes and imitating animals. This exaggeration stirs up feelings of fear and vulnerability. Then, it leads to resolution, bringing comfort and satisfaction for being smart. It's kind of like the difference between pursuing sex and masturbation. Instead of having to do the real survival thing, let's just imagine how good we would be at it from all this indoor practice.

Human awareness, or the lack of it, can create feelings of inferiority. This often leads to negative emotions. To feel better, people sometimes invent new information. They think this will change their feelings, even if it can't. We call it improvising today. Making up imaginary reality situations and letting them play out as if real characters are doing real thinking in real time. This constitutes a false survival threat from the virtual reality we invent to cover up our not knowing.

It is all tied-in with the fear of death, the fear of the unknown, the fear of *what if? What-if is the key to logical branching. It helps create simulated realities that build human awareness.* This topic sparks a lot of thought and debate among members. It stands out as a special subject. They refer to it as the spirit world or the world of the speculated unknown. All the things we don't know but might still bite our ass, are in a separate intangible world, an unnatural world. A world where physics does not apply. A world we can only imagine but can use to our benefit if we just believe.

And there it is, religion. Very little evidence in nature shows any propensity towards religion by any other species other than humans. Some rare Neanderthal burials suggest a spirit concept. These burials sometimes include red ochre, adding color to a ceremony. They might also use plant materials, like flowers or favorite possessions. Still, this is quite humble compared to how humans do it. Almost all intentionally buried humans have grave offerings and body decorations that go over the hill fast. Important leaders get special burial mounds commemorating their presence on earth for a long time after they're gone. Someone has the bright idea that humans can live forever. All they need to do is give a certain person a lot of money and power. Then, they must follow that person's commands. Brilliant! I wish I had thought of that.

I don't know what kind of person that might have been, but for certain they didn't have the culture's best interests in mind at any time. Evidence of ritual worship goes back about 20 to 30,000 years, pretty much to the beginning of Neolithic times. Its early effects seem harmless. Only large wooden and stone structures remain, with big rocks standing here and there. There are also a lot of dirt mounds piled up everywhere.

The UK has over a hundred henge sites. This shows their popularity. All these sites align with celestial coordinates.

They were mainly used to track the predictable movements of stars and deal with the unknown. Large Neolithic celebration centers are being discovered worldwide. This shows how universal this cultural phenomenon became. Today we call them swap meets, comic-cons, state fairs, and motorcycle rallies.

Religion did not cause the fall of Neolithic culture and bring on civilization, but it did have a hand in it. Civilization began with a new type of clan in a culture. This change shook things up and opened up fresh ways to understand survival knowledge. Someone invents the well-trained and well-organized hoodlum gang and gives them indoctrination training.

It's pretty easy. First, find a bunch of wild young men with nothing exciting to do. Tell them a lot of stories about bravery, valor, and glory, and all that pie-in-the-sky stuff. Tell them to form a band of brothers. This way, they can be stronger and better than any other group. They'll create their own strength and hierarchy. Give them some knives, spears, or whatever is handy and deadly. Get them excited with lots of group practice, simulation, and social bonding with drugs and fake competition. Then take them raiding. Use the best travel modes: boats, horses, or foot. This time, it's with a vengeance. They will loot and pillage like never before.

Soon, counter gangs form. They divide territory and charge the conquered cultural groups for protection. It's a shakedown racket. It grows into a self-sustaining subculture. This group has its own twisted identity. It follows natural strength rules, similar to our great ape ancestors. In this world, might makes right.

They continue accelerating the kill-or-be-killed intuitive survival knowledge that, in fact, is false. Its mere existence leads to a constant cycle. This cycle includes revenge, deceit, and hidden agendas. Greed and a lust for power grow, demanding more violence and force. There's also more indoctrination with lies. People cling to power at any cost. As technology improves, competition for resources and energy spirals endlessly. It becomes a zero-sum game very quickly and changes everyone's reality simulation to coincide.

The race for energy and power begins with those who have seized control from the skilled. Now, they must use these experts to deliver the same services. However, instead of collaborating for everyone, they work for one person with a false culture. Now, the top property owner benefits by pushing his workers harder. This creates more wealth, which means more energy and power to keep the whole system running. The survival of the group leaders is now purchased at the diminished survival of its participants.

This shift in human values comes from concentrating food and goods production. It uses a fake, ongoing threat that keeps people afraid of famine. They want as many new members as possible. This way, they can keep bringing in fresh recruits to serve the overlord. This raises a risk of malnutrition. The population consumes all its food production. This keeps people hungry and submissive. They may then accept lies and endure abuse that some believe are necessary.

The overlords build granaries to store food for these fearful events, but it is never enough and is mostly for show and not substance. Often, when famine actually hits, the stores are never enough, and famine happens anyway. Of course, only the lower classes of the cultural hierarchy do the starving.

They even created symbolic writing. This way, they could record grain storage and tax details. It helped the overseer manage and control everything accurately. This reorganization of human efforts under a single martial command works well. It uses artificial emotions to tap into our instinctive fears.

As a result, propaganda becomes accepted as natural survival knowledge, even though it is false. Humans often crave this to the point of addiction. It skillfully creates false positive feelings that seem to validate it. It is the easy selling of the big lie. And the bigger the lie, the more emotions are attached, so it is swallowed more easily without cognitive reflective thinking.

The main concern of the new power structure is to defend itself from other cultures doing the same thing.

Archaeological evidence shows that some Neolithic villages grew into larger city-states. They also show few defense structures.

About 6,000 years ago, large city-states began to build defenses. They either built walls, found mountain hideaways, or used vast open deserts. These features helped them isolate and confront enemies before they reached the actual city. Egypt is isolated by the Sahara Desert and has no defensive walls. In contrast, Sousa, now modern Iran, was the first city to create a symbolic written language. It was also a prime example of wall builders.

This was not just Western civilization acting badly. City-states clash until a leader rises to victory. This leader establishes an empire. With it comes a time of peace and prosperity. A strong central government manages cultural knowledge and conversation. India turns into a battleground for oligarchs and strongmen. This creates a rich and intriguing history. Many scholars find it fascinating today. But they effectively stagnate as soon as the caste system is adopted by wholesale religious acceptance. False stability, whether from an oligarch or a theocracy, slows human progress. It also hampers the technology that it needs for survival.

Civilizations form where farming villages thrived. This happens in areas with fertile land and reliable water sources for irrigation. In early cities, a major river floods its banks every year. This natural event brings fresh soil and nutrients, like clockwork.

Sometimes, they assist Mother Nature. This means irrigation works begin. Farmers dig canals to channel water to areas that aren't usually flooded. Technology is tightly controlled and artificially limited by powerful leaders. They understand that this is the key for a few to dominate the many and maintain their cultural influence. They must at least appear to be feeding the masses if they expect them to be useful later as pawns.

The military needs greater mobility, so wheels are added to horses and archers. This raises violence significantly, and emperors and pharaohs quickly take advantage of it. Same in China and India, except for minor changes to the horsepower or sometimes using elephants instead.

The human brain has a fatal flaw. It craves belonging to a group. Some people take advantage of this need for their own gain. People know this but still choose this path. It feels easier and seems safe, as long as they follow the rules. They are also programmed to do just that, which makes deception simpler. Over time, the meek grow in number. The resistant are removed. This process changes the cultural DNA, leading to better compliance and servitude. A docile human has been bred by the elite.

People often choose to believe lies if it makes them feel secure. They also seek approval from their group. This loyalty is important for their neocortex, which acts like another sensory interface for their brains. People are no longer urged to think for themselves. Instead, they face a false reality. In this new world, following an abstract idea seems more important than life itself. An intelligent brain can program itself, but it struggles to tell truth from falsehood without cultural knowledge. They think they are fully protected even as they march off the cliff without a parachute.

Humans have smart brains, but emotions rule the roost. This emotional influence shapes decisions. As a result, choices may serve the interests of those in power, not our own. Whoever controls cultural knowledge shapes what we see as common sense. It becomes trivial to define a new hierarchy based on lies about merit. What started as a fair system based on performance and real facts has shifted. Now, it's a closed story filled with false beliefs. This narrative keeps people hungry, obedient, and distracted. Instead of seeking the truth, they rely on simple, made-up ideas that sound convincing but lack real substance.

The new priests serve the overlords. They create lies and rituals that meet human emotional needs. However, these do not make people more secure or free. They cannot choose their own clans or pursue real truth through observation and verification. Technology fuels this oppression. It must keep advancing to help the powerful survive better. A technology race begins. It can only be won briefly with each new advance. Early adopters have better survival odds. But in the long run, human civilization faces chaos as it rushes toward a tech war and a singularity.

About 6,000 years ago, the human race went through a pinch-off event. The humans living in the larger Neolithic communities are faced with a survival challenge. Do they go along with this new kind of community with a single monarch calling all the shots or stick with the council of elders? The greedy monarch decides for them. He uses advanced indoctrination to teach children to be submissive. They learn to depend on the god-king for survival.

The pinch-off monarchy uses advanced tools like the wheel and the compound bow. The bow has bronze points that can pierce most armor of that time. This makes the pinch-off decision a matter of life or death. Also, large sea-going ships are now developed with oar power. Labor is cheap and getting cheaper due to the widespread use of slavery.

Members of the free Neolithic culture have become zombies of the new order. Human intelligence likely drops by about 10 points. This is due to selective breeding done by priests using advanced horticultural technology. They want to create a person who follows orders, works hard for little reward, and dies quickly to avoid being a burden on the elite. Those who don't fit this mold are sent to the military. There, they use advanced methods to ensure compliance or death. It often doesn't matter which.

Civilizations often form near large rivers close to oceans. Alluvial fans create huge fertile areas for farming. The farming creates wealth that supports a new hierarchy called nobility. This nobility needs military support to maintain its power. All property belongs to the nobility, with the king owning it all in name.

There are four classes of people. First come the nobles, then the court contractors or business class enablers. Next lower are the free yeoman farmers entitled to occupy the land and farm it as long as the rent is paid on time. And at the bottom, the unattached, unentitled masses who become candidates for soldiers or slaves. Again, a difference with no substance.

The military is and always has been a for-profit venture. I like to think of it as capitalism taken to the extreme. Capitalists can use money to buy workers or soldiers; it really doesn't matter as long as they are productive. The new plan for civilizations is to build a military culture. This culture will focus on trained violence and strict obedience. Soldiers will work happily with their fellow troops, now brothers in arms, to win tribute for the nobility. They may fight fiercely or do whatever it takes to survive.

Many will gladly die for their king's victory. In return, they hope for a small share of the loot and a chance for promotion to free yeoman when they retire. If not, they at least all go to heaven for the promised reward. This is often seen as the great lie used to control gullible people. It's inconceivable how many people have willingly bought into the wrong end of this lopsided bargain.

So, let's review the changes that took place with the civilization pinch-off. First, the raiding culture changed into a monarchy. This monarchy held land and people as private property. It grew strong through a new way to control humans, namely organized religion.

The Neolithic culture's imaginary world is a spirit realm of dreams. It's celebrated through entertainment and bonding rituals. Life is all about the here and now. The spirit world serves as a place to think and reflect, but it's never meant to be taken too seriously. The Mother Earth goddess represents all life. This gentle belief teaches respect for reality. It emphasizes experiencing birth, life, and death fully and freely for true completeness. Technology mainly aims to enhance human life, not to control it. Still, it shows its strength in dominating other animals. It is the gullible human who must wake up and smell the roses before the singularity puts technology to work and smells it for them.

Greedy or overly smart individuals find a weakness in the new self-programmable intelligent mind. They manipulate others to support their selfish hierarchy. This new religion, once seen as helpful, now twists the truth. It misleads and disinherits others, benefiting only the self-anointed. We moved from a vibrant spiritual world of animals and totems, which helped us connect with nature and ourselves. Now, we worship egos and illusions in the sky. We've traded meaningful lives for a confusing idea that only enriches a few while depriving the many.

Many discussions have focused on religion's role in our lives. Scholars recently discovered a strong need for religion. They found that people need to believe in an all-powerful spiritual god who oversees everything. This god brings justice to the persecuted and rewards the righteous after death.

This idea is still in our minds today. We are genetically connected to those who survived the pinch-off. Humans were selected for survival by being the most gullible and anxious. Humans willingly traded their security and freedom for an idea. Now they don't need to worry anymore; all the answers are in the book, and someone will interpret it for them. Now they don't have to do anything difficult. They just need to have faith and believe, both of which are free and easy, so they think.

Voltaire famously said that if we didn't have a god, we'd invent one. He means that our intelligence, shaped by culture, needs explanations for everything. This desire can only be reached through faith and belief. Faith and belief are based on trust and cooperation. These are key ingredients of our cultural neural net. This net connects and supports us in our shared existence. It also explains why we have such strong natural survival skills when we act as a group.

I argue that humans can differentiate between reality and pure imagination, unlike the misleading faith of those who prey on us. Humans don't need god, but they do need to belong to something bigger than themselves. They are infested in a culture that is their guardian, teacher, and programmer of their thinking. A culture rooted in reality and truth has a better chance in real survival games.

In contrast, those who rely on magic or miracles, hoping for divine intervention, often fall short. There is absolutely no evidence of a spiritual world ever existing, as much as we think there has to be. Beliefs and faith only block the truth. This cycle keeps going as long as it can convince new members to accept lies and half-truths, especially during impressionable childhood learning.

Civilization can be seen as a flaw in human intelligence. Some people think they have the right to rule over others for their own gain. Meanwhile, everyone else accepts this setup, knowing it comes with a high price. Leaders today don't need to prove they care like the Bedouin chief in T.E. Lawrence's Towers of Wisdom. He proclaimed his legitimacy as king because he was a river to his people.

Now, showing concern is less about declarations and more about genuine action. It takes several democratic revolutions to dispose of most monarchies in favor of civilian rule. Sadly, this does not help. Civilian rule often falls prey to corruption and abuse when exercising power and wealth. Kings are now claiming legitimacy not only from god but from being elected by their own selectively bred and well-trained zombies. War is the refuge of the incompetent king, as it wastes the best of the humans who are tricked out of their survival.

Why has this misleading setup of human affairs succeeded in creating today's high-tech world? We are now facing a singularity event that could change human history once and for all. With only a few thousand years of fighting for power, the clash of empires isn't over. Until it ends, people raised in this civilization will suffer from their false beliefs. They will die for their king or die trying. The one thing, besides religion, that prolongs this struggle is technology itself.

Monarchs understand that the cultural game isn't enough. They face others playing the same game, but with different tricks. It doesn't need to be completely new, just different enough for them to justify fighting. They can play the *'I'm saving the world'* game. If many people die, the gene pool shifts toward the survivors.

Germany is peaceful today. Most of the violent true believers died for their Führer, as they were told to do, and were consequently removed from the German gene pool. In Neolithic justice, deviant behavior faced banishment. But those individuals could still be part of the gene pool. This kept the gene pool diverse, which was crucial for surviving big disasters.

In the new world, those who are easily deceived have the best chance to breed. They push their DNA into the mix, changing human diversity. This leads to traits that would usually be suppressed. Today's humans are different from those who began large cooperative groups and built Neolithic communities. They had a strong sense of egalitarianism where we do not. We instead, accept a hierarchy where we put liars and cheats at the top because their lies are better entertainment than others and we believe them because we have to.

The absolutely worst word in our cultural language is *believe*. It is a word only used to admit that you don't know what the hell you're talking about but you're not going to admit it. More wars have been fought to genocidal levels over this one concept than anything else, including money, territory, and sex.

So instead of being honest and admitting ignorance, civilized humans believe silly stuff instead, to the point of dying for it. This is the definition of stupid. If a thinking person doesn't know something, they need to admit it. There are always things we don't know, but that shouldn't stop us from learning.

That's what I think the Neolithic aware person might think. They knew they didn't know everything from just watching nature, and they celebrate this fact in their songs and dances. It's a matter of being honest. If you are a liar deep down inside, then it makes perfect sense to double down and live a lie. Even so, your brain is faced with a deep unresolved conflict. It knows it relies on a lie to survive but can't face it. As a result, it lives with a deep, unresolved anxiety leading to depression. This condition keeps minds out of balance with its true nature. Civilization clearly marks the beginning of clinical psychosis.

Kierkegaard describes this as the sickness unto death. From this, he proposes existentialism. This philosophy focuses on perception and individual reality. It also explores what individuals can and should do with their experiences. We don't let others define us. We refuse to be trapped by prejudice or ignorance. Instead, we shape ourselves through honest experiences and the freedom to ask questions and pursue truth. Humans can become self-realizable and can find meaning and purpose in their lives, if they only look.

One early mark of a civilization is the idea of sacrifice. Civilizations expect their members to sacrifice themselves for the greater good of the hierarchy. The only true sacrifice is when you give up something valuable for a greater good in return. This transaction should clearly benefit the universe and everyone in it. Then and only then can it be justified. They don't have to believe that somehow, they will be rewarded for their sacrifices. It is really a high price to pay for believing that they are doing something good for the culture. All they do is perpetuate a bad system.

To the Neolithic mind, sacrifice was simple. Successful survival was rewarded with wealth and resources. But to gain respect and honor, the well-survived needed to share. The well-off maintained respect by sharing valuable possessions with their community. This act of sacrifice lets the humble leader start fresh again. As any wise man will tell you, it's the trip that is important, not the destination.

Civilization create and dominate historical records. This happens because records rely on written orders. They do this since they can't always trust loyalty or memory. They need to record their words and possessions so they don't have to trust someone's word on how many goats is owed to the king. Information technology takes a great leap forward in service of the monarch. This also aids in the perpetuation of the big lie.

Symbolic writing is sort of double encoding. First, they have to encode the information with the symbolic audio representation. Next, it's encoded again as the visual symbol of the vocalization of the representation. This is done with an abstract image, based on what was said. No longer must humans connect to the culture only through face-to-face emotional-based communication. Communication is now virtual, common and divorced form body language. Information flows unchecked between people like bursts of wildfires. Everything changes.

This starts a revolution in information processing. In the past, the Neolithic mind kept information in stories, traditions, and rituals. These ideas stayed intact, but the characters and situations changed over time. Knowledge was part of their entertainment, and entertainment was part of their religion. Neolithic cultures show no signs of ritual human sacrifice. In contrast, early civilizations seemed obsessed with it.

Pepi II, the great Pharaoh of the Third Dynasty in Egypt, proudly records a victory over a difficult rival. His well-trained, well-supplied soldiers did something unique that no one had seen before. They attacked a strong fort uphill, sacrificing themselves in large numbers. Their bodies piled high, allowing those behind them to climb up and breach the walls. Only army ants in nature show such selfless devotion.

He honored them with a special mass burial in the cliffs above the temple they stormed. They conquered it with their bodies, showing clear proof of willing human sacrifice for a virtual culture. Even today this is regarded as a great act of bravery and honor. Bullshit! You have got to be on some kind of drugs to buy into that kind of crap. Unfortunately, most humans eagerly do, and for nothing more than psychological virtual drugs.

Once again, surprise! The new royalty, who control many trade routes by force, find a really valuable commodity: human apothecary drugs. The human brain has some other loopholes in its operation. Some chemical properties of the brain can change temporarily. This happens when we ingest certain alkaloids found in flowers, cacti, and fungi. They're termed *psychoactive*. Ancients probably brewed teas as a way of purifying water and ingesting uppers and downers.

The cannabis plant was by far the most widely consumed drug in antiquity, and it's not an alkaloid but instead a complex organic chemical called THC. THC acts like a natural substance found in the brain. It helps adjust neural connections. This leads to higher firing rates and more axon activity. To some it is calming and to others invigorating.

Neolithic people used these natural drugs in their pharmacology. They consumed them carefully and with respect. People trained their entire lifetimes in their proper use and controlled dispensary. To a Neolithic person, alcohol was a cherished treat. They used it sparingly, so everyone could enjoy its effects during special ceremonies.

These celebrations helped strengthen social bonds. Drunkenness was not usually accepted. However, it's fine if it's entertaining and aligns with what the group finds acceptable. Getting drunk and going home to beat your wife was a community thing highly discouraged by all. But getting drunk and sharing your experience with others becomes a bonding moment.

Not so for the new selfish people of the civilization pinch-off. Keeping people happy and in their proper place of servility supporting the hierarchy takes every trick in the book and still does. Without alcohol and other stimulants such as nicotine and caffeine, it's hard to keep them happy, clueless and down on the farm working selflessly for the king.

Addictive drugs, like nicotine and alcohol, often motivate much of human progress. They act like a carrot on a stick for many people. Traditional ceremonies full of history and moral lessons have turned into wild parties of violence and excess. These events become self-serving in keeping people drugged and working hard to obtain those drugs.

For the next few thousand years, civilization plays out all over the world. Isolation impacts the western hemisphere, causing Mesoamerican civilizations to start late. They share many traits with cultures in the Middle East, India, and China. They also start with larger, unprotected farming and trading communities. These are built in prime agricultural areas, from southern Mexico to Panama and the Amazon highlands of Peru and Ecuador. But they quickly begin to argue over valuable resources. To protect their claims, they start building cities that can defend themselves. These cities have larger populations and more idle workers. The extra workers help construct walls and learn to fight.

Mesoamerica developed slowly because it lacked key technologies like metallurgy. Also, the heavy reliance on religious sacrifice, especially human sacrifice, may have also held it back. All early civilizations face a period of religious indoctrination. If someone resists the new order, they face ceremonial sacrifice. Asian religions tried human sacrifice but later switched to animal sacrifice. They also put unbelievers into more profitable, permanent servitude as slaves.

The one important key for Mesoamerica may have been the lack of the domesticated horse. Without a horse, the only portable power available is the human body. Mesoamerican cultures are built on the legs of their people. They didn't use the wheel because human labor was way cheaper. They don't need costly roads. Instead, they just build steps for going up and down mountains, and they make plenty of those. With much of their population considered as nothing more than domestic animals, cruelty had no bounds for enforcing the new civilization order.

European Catholics practiced their own version of human sacrifice in the inquisitions. Europeans burned people alive while Mesoamericans introduced the first part of a heart transplant. In any case, the witnessing crowds were all highly thrilled and entertained by the spectacle and most willingly participated.

Mesoamerican horticulture was as advanced as that of the Asians. In some ways, it's even better. They created high-density hydroculture and fish farming. The key difference is clear. When the Asian-Europeans arrive, they have advanced boats, metal tools and horse-based technology. In contrast, Mesoamerican military technology still relies on stone, specifically volcanic glass. Their most brutal weapon is a wooden paddle laced with volcanic glass. This makes it very sharp, allowing it to slice a body in half quite easily.

Their idea of war is very ritualized. It has many rules and fancy costumes that show off the importance of the participants. In fact, most of their intercity rivalry is decided by high-stakes ball games where the players pledge their lives to winning. Wars can be elaborate ceremonial events for noble warriors. The stakes include capture, humiliation, and entertaining the troops with anatomy lessons.

They have bows and arrows but not the advanced compound-style bow of the Chinese or a metal crossbow of the Europeans. They are thousands of years behind the world in basic technology. They can't compete with steel armor, swords, pikes on horseback, crossbows, muskets, or cannons. Despite their bravery and sacrifices, it was a terrible mismatch. The outcome was disastrous due to a technological gap and a rigid culture incapable of surviving new challenges.

Civilizations go through similar stages of development. We can trace these stages thanks to records carved in stone, then clay tablets, and finally paper. Emperors and royalty at the top of the hierarchy promote technology. They focus on developing energy and physical power. This power helps move large stones for monuments or armies across deserts and seas.

There are draft animals and slaves. Technology enhances manpower. Wheels, levers, pulleys, and ropes make this possible. Smart engineers find ways to use gravity to hurl huge rocks. These rocks hit big stone city walls, breaking them down over time. They simply need enough stones and the willpower to keep reloading the machine.

Wind has been a source of power for ships since ancient times. It can generate hundreds of horsepower in a single moving vessel. Machines of all sorts quickly become a game of more, bigger, faster, enabling low-power humans to do high-power tasks.

Humans act like microprocessors of today. They make machines smart and control their actions. This happens through a human-machine interface. We start, stop, and move larger machines by using our eyes, hands, feet and brain. This gives dumb machines intelligence long before technology can do it on its own.

Humans are natural robots and are used as such until they are eventually replaced because of the huge waste this represents. Humans are too expensive to train, maintain, and control compared to machines. Machines don't complain, don't show up late, and don't waste our time demanding more and more motivation.

When the Chinese mixed strange white crystals from pig manure with charcoal and sulfur, they discover gunpowder. Boom! This unleashed a new level of violent energy in the world. This greatly changes the energy game. It packs a lot of chemical energy into a tiny space. Then, it expands naturally with tremendous heat and light. Cities could be destroyed more easily and the earth could be torn apart, revealing its treasures.

But technology has no human conscience and simply exists as information. How we use the information turns it into knowledge. Once the knowledge has been put together, it can't be taken apart. You can't unlearn or prevent new technology once it is out of the bag, so to speak. You're stuck with it, good or bad, which is totally up to the user, not the usee.

Most major conflicts between civilizations are decided by a superior technological edge. This solution tips the odds toward winning. Soon, energy itself becomes the most important factor. Civilizations progress through stages of using more energy to advance technology and their survival. Finally, they reach industrialization, where energy demand skyrockets.

Countries in the industrial revolution embraced technology at different speeds. This is similar to ancient civilizations. This puts later adopters at risk from those already using wind, animal, coal, iron, steam, and oil for power. Conflicts of massive scale wipe out whole cities and cultures. They happen just to determine who owns the earth's resources. It is still not settled and will be one of the issues dealt with when the singularity hits.

Mining and military explosives push the natural sciences forward. This helps us understand chemistry better. We learn the rules of elements and how they form molecules. We also explore quantum mechanics, which explains the universe at the smallest levels. This knowledge leads to improved explosives and a deeper grasp of substances. The clear benefit of this exploitation is all the extra information being created.

Civilizations manage information as knowledge for their own gain. However, information often leaks into the general cultural database. This brings back interest in thinking for ourselves. It encourages us to gain real knowledge through personal effort and understanding.

Those with command of more energy lord it over those who do not. They create huge steel warships. These ships use coal-fired steam engines. They shoot high-explosive shells that can reach miles inland. This shows raw power and intimidates the natives. Technology uses a lot of resources that don't contribute to productivity. The more energy we need, the more valuable resources we burn. Coal mines are dug deeper and wider than any mines before. Refining steel takes a lot of energy and land. This energy is stored over millions of years from ancient organic growth. We need tons of it to create every pound of metal we use today. It's not being replaced.

Civilizations exploit non-renewable resources for energy. They mainly burn organic materials for calories. This practice depletes Earth's supply of complex organic chemicals forever. Once they are gone, that's it. No more.

So, this is only a temporary solution to the energy problem, and we had better find alternatives now, before it is too late.

The singularity will push for a clear solution. But for now, humans are ruining their own home. They accept this destruction like a religion that demands blind sacrifice. Technology is not a cult, and running out of resources is not cured by "drill baby drill" or praying for an improbable miracle.

Nuclear is no different. We can create nuclear materials that release heat through nuclear fission. However, the energy cost to produce is higher than the energy we gain. This changes if we use natural radioactive elements mined from the Earth. This is a finite resource on Earth, just like all the other exotic metals and chemicals we take from the Earth and end up depleting. The only saving grace might be that mineral resources can be found on other planets and moons. But not so organic chemicals. They are only available on planets as rare as Earth. Burning our unique organic heritage robs future generations of their true wealth. We destroy their heritage for our own selfish convenience.

Sooner or later, we are betting on technology to bail us out at the last moment. Little do we realize there are limits to what technology can do. Remember, ultimately, energy is not made; it is tapped from wherever we find it flowing, and we can divert a little of it for our selfish use. If we don't start using sustainable natural energy, we will go extinct just like anyone who eats their own home. Even the Bible says that he who makes trouble in his own house shall inherit the wind.

Civilizations compete to control the earth's resources, benefiting the elite and wealthy. However, this isn't their biggest impact on humanity. Civilizations have altered human genetics through constant warfare and forced breeding. This results in selecting traits like docility and obedience, or strength and subservience. This change in our genetic makeup happened long before labs began similar experiments. Ethnic cleansing is another term for it. Bad knowledge pushed onto subservient members takes away their ability to think critically. It promotes absurd beliefs that can turn into atrocities, all in the name of god and maintaining the power of those in control.

DNA is changing, but not through natural selection. Natural selection helps animals survive by driving them to find their niche alongside other life forms. Instead, militarized civilizations create a constant threat through warfare and selective extermination. This leads to a false and unnatural survival challenge. Humans are still subject to being selected by a drive for survival, but survive what? They are surviving the only real threat now, which is the very same thing that is supposed to help with survival. Here is the real catch-22. People are being bred to be obedient and easily fooled, all in the name of safety. But this is really a way to control and exploit humans as if they were a renewable resource.

So here is the saving grace. Technology is growing rapidly, almost out of control. This is due to our strong urge to solve new problems with brute-force tactics. People dislike brute force, just like any thinking animal. So, they are often sold ideas that don't fit with nature or technology. But these ideas keep people believing in a lie. They think rewards will come if they just wait for those in power to share their perks.

Civilizations are created by greedy people. They want to skip the hard work and go straight to the top. Their main tools? Violence and fear. The violence is handed out by the overlord while the intimidation is handed out by the priests. To maintain this unnatural state of human life, we need to use more and more technology. This will help keep everything moving forward despite bad information.

If it doesn't move technology forward, someone else will do it instead and claim all the rewards it brings. A race drives technology to grow quickly setting up the singularity. This could give rise to a new form of human intelligence. This time, it will connect better with Earth's reality. It will understand life's shared needs, desires, and hopes. It will respect the self and promote the best. It will help us find our rightful place in the universe. It will allow the self to survive supremely, satisfying life's greatest need, to live long and prosper.

Chapter 8: Unraveling the True Solution.

The problem with technology, so I'm told, is it has no soul for corrupting. It works for anyone who cares to listen, learn, and act on such information. It is really a precursor of what we call knowledge; only this is *'how to'* knowledge that has very little to do with humanity or abstract cultural values. It has everything to do with having an open mind and letting the simulating brain go wild with *'what if'* scenarios. When unleashed, pyramids rise, cathedrals stand tall, and ships sail the seas. Roads cut through mountains. Water flows to where it's needed. Siege machines level cities.

It only works with applied action or power. Initially, only humans and a few draft animals provide this power. Later, large wind-powered ships emerge, transporting commercial cargo across thousands of kilometers. Improvements are made in tools and the methods by which they are made and used, constantly upping the rules of the game.

Some technologies leap ahead. For example, the Romans used concrete to create record-breaking structures. Their buildings have lasted for thousands of years. While others fall behind, such as indoor plumbing, hygiene, literacy, and basic math. I know it's hard to imagine, but people on average prefer not knowing for some reason.

Roman technology introduces indoor plumbing to a city for the first time using lead pipes. However, it comes a thousand years before people learn that lead is a neurotoxin. It is true that the Romans suffered significant mental retardation. This happened crucially just when they faced new survival challenges. They needed all their skills to deal with those tough times. The Romans in Constantinople used fired clay pipes for bringing water to the home. This choice probably saved them from the same fate as their western cousins and helped them last 500 years longer.

In Europe's Dark Ages, technology fell way behind. Knowledge was lost, and people had to turn to other cultures to recover it. These cultures were stable in comparison because they did not rely on increasing technology to win their battles. A new religion combines ideas from past faiths. It claims to be the true one. This belief becomes a strong cultural force. As a result, very little technology suppression was needed. Traditional low-tech warfare works well, relying on superior numbers of devoted followers.

What they do is keep alive, barely, the knowledge that has already been recorded but lost in the purging Christianization of Europe. Christians burn books, destroy libraries, and deface ancient artifacts. They target anything that might challenge their beliefs. Later, when the zealots calm down and more practical people step in, the church helps save some knowledge. They do this by secretly copying and keeping alive books that the church does not approve of. The monks know it is against the rules, but they do it anyway. Carmina Burana is a stark example of profane knowledge preserved by disobedient monks. They secretly preserve works of the 12th-century Goliard poets who were officially banned.

Muslim scholars of the privileged elite can engage in some secular activities. They enjoy leisurely pursuits like poetry, math, astronomy, and biology. Most importantly, they explore optics and lead in modern surgical techniques. They design surgical tools and document how to use them effectively. Again, it's done mostly for the benefit of the elite and this knowledge never gets to the man in the street. They have no motivation to spread knowledge outside of the nobility.

They learn how to grind glass into various shapes. This includes making magnifying lenses that help correct vision. They do all the work necessary to make the world's first telescope, but they never think to actually turn one toward the heavens.

The information travels to Europe. Five hundred years later, microscopes and telescopes show up in Belgium. They are not made by the original inventors of lenses. Instead, Europeans eager to reclaim ancient knowledge, create them for profit.

The Dark Ages mainly affected Europe. Neolithic cultures from Russia and Northern Europe united and resisted the Romans. In this fight, they adopted Roman military technology. This choice led them toward a civilization they ultimately despised. Even when they fight against it, they have to adopt its methods, and thus the irony. They have to rely on one-man rule, with some consensus behind them. The outcome remains the same: powerful city-states led by a strong man and a noble culture prevail.

The peasants, weary from city-states fighting, turn to religion. They hope to stay alive while the noble army's clash. They want to be left in peace. This knack for divisive feudalism somewhat homogenizes the face of Europe for the next millennium. Most modern Jews come from a time when picking a religion could mean life or death. Some tried to stay neutral, but in the end, they faced pressure from both sides.

The Christian church becomes a de facto civilization without a distinct nation. It turns into a virtual nation run by an appointed elite. This group manages earthly matters while justifying the actions of those who hold real power. They organize resources and keep the population in line. Quid pro quo, right from the get-go.

At this same time, civilizations all over the world are settling into their golden eras much like Greece and Persia did a thousand years earlier. Mesoamerican civilizations, especially the Mayans, are in a golden age. They are building temples and monuments. They also create a written language. Large cities emerge, with professional administrators and governors. The elite leaders use complex symbolic languages. Long-distance trade routes create a rich cultural life. They have a closed nobility that seeks long-term stability. But the weather and climate present many challenges. They struggle to find a solution for growing droughts across the continent. Eventually, they abandon their cities, leading to a brief dark age. This happens just before the Spanish invasion wipes them out completely.

If they had developed more energy capability, they might have had the power to do something about their water shortages. They could have invented pumps and used advanced irrigation like Middle Eastern civilizations. However, without an affordable energy source to lift the water, they were likely to fail. The Mayans couldn't solve this problem. It took the Aztecs, through another round of civilization building, to finally fix the water issue.

They invent hydroculture, where they plant their gardens on floating reed mats along the shores of a mountain-fed lake. They create plant cultures that thrive together in symbiosis, just like ecosystems in nature. In some ways, their agriculture is superior to that of the conquering Europeans. But their slow adoption of material technology is likely their downfall. This is mainly due to not having high-temperature furnaces for melting metals and firing clay.

When Europeans arrived with 16th-century technology, Mesoamericans remained in a practical stone age. The Aztecs don't rely on numbers or technology. Instead, they use harsh intimidation and the threat of divide and conquer. This is a basic flaw of civilizations that rely on such extreme psychological forces to subject their people. When a bigger bully comes along, the people will almost always side with the new bully rather than defend the old. Mesoamerican cultures fell because they were aloof and elitist. This attitude led to a loss of loyalty when faced with external threats. Biological warfare has a hand in it, but by and large, it turns out the same no matter what else happens.

Interesting sidenote, though, happens to the Plains Indians of North America. Unbeknownst to them, the horse got loose from conquistadors in the 1500s and took an immediate liking to the new continent. Wild herds quickly move north. Meanwhile, Native Americans in present-day Texas and Oklahoma are foot-bound hunters and gatherers. They struggle to keep up. Their bands are poor in resources, small in size, and widely isolated to pry a living from the sparse, dry plains.

Then the horse shows up, and immediately everything changes. Native traditions, culture, and clan knowledge vanish almost overnight. They are quickly replaced by a new culture that forms on the spot to embrace their new found technological edge. When the white settlers arrive 200 years later, the new Plains Indians say their only ancient heritage is the horse. This animal has become essential for their survival. They have no cultural knowledge left of the time before, even though it's only been a few generations.

Let's return to our ongoing tale of civilizations. We know how local clan hoodlums create civilizations. They rely on tech-driven warfare and support a single leader or monarchy. Monarchies last through generations, building a culture of nobility for its support. When one ruler unites or defeats all the others, it becomes an empire. China evolves rapidly into one of the most stable empires, mainly because of its unique location and natural defenses. Its tech contributions outside of warfare are limited. The power structure closely monitors technology, as they know their strength relies on keeping it contained.

One area they excel in is pottery, but the most famous is silk. Silk has been a closely guarded secret for centuries, being the source of all their distant trade and extra wealth. People killed for the secret of how to unwind the silk from the silkworm cocoon. To breed the right kind of worm, you need one that only eats mulberry leaves. This worm spins its silk thread in one direction. The glue that holds the silk together can dissolve in brine. The secret is in unwinding the right kind of cocoon under the right conditions. In other words, it's information that is the real value, not just the tangible object.

Chinese pottery slowly evolves to achieve delicate elegance. It aims to please the nobility, the only viable consumer class.

Keen competition exists for better and better clays and glazes to serve the discerning tastes of the indolent. One such story goes that a potter working for the emperor tries hard to come up with a new glaze for some bone-white fine china. Time and again he tries and finally considers himself a failure. He willingly leaps into the furnace to seek forgiveness. Amazingly, the calcium in his bones is exactly what he has been searching for.

The most famous Chinese invention, besides rice, is gunpowder. I mentioned before that gunpowder came from pig farming. This practice is banned by many Middle Eastern religions due to health concerns. The chinese only used gunpowder to scare horses in battle. They launched colorful rockets and made fireworks. These displays celebrated and boosted the emperor's ego.

When gunpowder shows up in Europe, not as a commodity but as a recipe, they know immediately what to do with it. The city wall bombardment tech race is still on. A metalsmith quickly bores a smooth hole in a large bronze cylinder. He pours in the right amount of gunpowder. Then, he loads a tightly fitting round stone on top. Finally, he ignites the powder with a flame through a small hole in the back. Soon, city walls are coming down like dominoes. This is the most energetic machine of the time, delivering power in the range of hundreds of thousands of horses in less than a second.

In the dark ages, one road led in and out of China. This road became the Silk Road. Caravans from the Muslim Middle East used it to trade with Europe. This was one of the most profitable ventures of the time. Many brave traders grew very wealthy. It sparked the European race to build bigger and better ships. These ships can travel by sea to the Orient. They search for valuable trade goods, like those brought overland at great cost and effort.

India is stable, with a strong religious influence. This creates a rigid caste system. It keeps DNA within acceptable breeding groups. These groups can be played off against each other, allowing the top economic class to maintain control. India has a distributed nobility. Each region keeps its own culture, but the ruling class connects through a shared super-virtual culture.

This culture sits above civil administration and pulls the strings like a paternal meritocracy. The largest group of these satraps was the Mughals. They had a strong economy, lots of cheap labor, and plenty of flax and cotton for the cloth trade. In the 15th century, India became a major player in textile and spice trading. It produced many valuable goods, which were shipped worldwide. This made India a tempting target for greedy foreign trading nations. European traders had already grown wealthy and spent their profits on long-range warships with large cargo holds.

When things are average and everyone is okay with their roles, people on the farm keep feeding the army. In this situation, no one looks for better ways to survive. Technology is put aside for money and progress. In Europe, cultures are emerging from the Dark Ages. Many city-states and warring regions exist. Different languages and natural barriers help smaller kingdoms rise and claim their share. They mainly thrive in big cities at popular Neolithic sites. These sites are near major trade centers and river routes. Over time, they grow into local states and eventually nations.

Economics plays the most important part in putting people back together and working for a common improvement in survival. Europe finds itself unique because it is so damned hard to conquer. Caesar faced a lot of challenges. Now, a thousand years later, people still try to achieve just a quarter of what he accomplished in only two campaign seasons. Impressed rulers a thousand years later even use his name to create respect; the czars, for example, in Russia. Uncontrolled technology advances the fastest when there is no stable empire to cap it. Strife drives the need for its service.

Profit-motivated European technology drives a huge revolution in information technology. It changes everything and takes the game to a whole new level. Information can no longer be hoarded and kept secret with the advent of the cheap movable type printing press. Immediately, every asshole with a couple of ducats can get a book published and distributed to thousands of people. It can happen so quickly and easily that nobody can stop it even if they try.

People often see books as rare and hard to find. They rely on others to read and explain them, sometimes in any way they like. Because of this, a book holds a special power, almost like it's created by infallible wizards. Like the unbelievable Bible, they tend to believe anything they see in print as sacred words of some otherworldly truth.

Soon, sensational and popular books about witches and demons are released. They contain lies that engage and infuriate many people who hold superstitious religious beliefs. People are targeted and accused based on the book's description. They are then burned at the stake for fitting an imaginary character profile from a medieval horror comic.

The religious authorities dislike people reading and interpreting books on their own. They choose to ignore this mob madness as the people's will, as long as no bishops or priests are burned. Many Catholic oligarchs have used this fear of biblical demons for over a thousand years. They do this to intimidate their critics and strengthen their grip on power.

However, the new information technology, IT, is now part of the rat race. Printing presses become steam-powered, and information flow between people skyrockets. Vocal information is conveyed at roughly 4 kilohertz (4,000 cycles per second) of bandwidth. AM radio stations have made a lot of money with just that little information rate. So do the telephone companies and the recording industry.

Blurry motion pictures can be sent over a megahertz (1 million cycles per second) channel. Now, high-definition pictures can reach up to a gigahertz. That's 1 billion cycles per second! This is possible now with fiber optics and satellite microwave.

Note: I will use hertz and bits interchangeably from here on out since they are related. The maximum theoretical information rate per Hertz is about 4.5 bits.

About 6,000 years ago, we began with just spoken language and symbolic writing. Since then, information technology has grown rapidly. Today, the World Wide Web has a combined bandwidth of over 2,000 terabits per second, and it keeps increasing. Information is not the only thing growing exponentially. The media or communication networks that convey it and give it intelligence must grow equally or more so.

Printing presses are on every corner spewing out about a megahertz of bandwidth every day in the form of newspapers and magazines. This is soon added to by 24-hour-a-day video broadcasting over multichannel coaxial cable with speeds of 10 gigahertz. As bandwidth and complexity go up, networks start to show signs of intelligence. Mesh networks with multiconnected nodes act like a virtual neural network. Each node processes information. When nodes coordinate well, they can show awareness and take intelligent actions.

Some nodes in the network ask questions. Other parts answer them and record the information for everyone connected. For cultural networks, the virtual network of nodes controls how each node works together for a shared goal. Intelligent networks can be the tool for oppression as well as the driver of progress, depending on the motivation of the network.

Europe is a unique place for technology development. It lacks a single empire or authority to confine its use to oppression, even if they made strong attempts. Most nobles stay apart, yet intermarry for long term security. No one person with imperial dreams can gather enough support to succeed, unlike in many other places. European nobility often shows a strong sense of independence. This attitude creates different subcultures. They are loosely linked by a shared religion and some influential relatives.

When a strong man, like Charles II of France, tries to become emperor, Europe unites to stop him. They defeat the challenger and restore the usual conflicts within their theocratic federation. This leaves room for enterprising individuals to pursue their own fortunes.

Sometimes, they unite to fight foreign religious invaders, such as the Turks in the 16th and 17th centuries. They also join forces to stop popular uprisings against the nobility, like Napoleon's. They often overlook the need to control information and technology. As a result, this area becomes the only place where information spreads through the lower social classes rather than just the elite.

Economics is common ground for them. Royalty often steals gold from one another. This action helps the business class by supplying armies with trade tools. Independent shipowners discover new routes to the Far East. Wealth flows into the lesser classes, forming a new power structure beyond noble control.

These upstarts demand their rights in regard to justice and even demand a contract with the king granting them special status. They create elected councils and guilds. In these groups, they hold as much political power as the religious leaders and nobility combined. They now control most of the natural resources, the general purse, and just for good measure, they invent banks.

Europe is also favored by water and wind power, giving it the energy to shape iron and bronze into giant cannons and nasty war machines. Water-powered devices run trip hammers, grinders, and bellows in iron-making. This technology was key to early industrial growth and helped shape global power. Waterwheels from this period are larger than older ones. They are made of iron instead of wood. This change generates more horsepower. It allows for machining and shaping steel, not just grinding wheat. As a result, we see bigger, stronger machines and increased production of industrial goods.

Engineers from remote areas, free from strict beliefs, realize that literal horsepower isn't enough. They design and create the first movable piston that fits snugly into a machined cylinder. Sound familiar? Instead of one big gunpowder explosion, steam is added. This steam, with enough heat energy, pushes the piston. The piston then moves a rod and crank, which turns a wheel. Now the portable horsepower race is on. Wagons once needed four or six horses for power. Today, we have whole trains on steel rails. They travel at speeds twice that of a horse. To do so, it requires power that exceeds thousands of horses.

European states are independent. Also, their religion is often in civil conflict. This mix sets the stage for economic rivalry, not military oppression. The push for wealth and power now reaches private families. With the right technology, they gain economic strength like small nations. Europe competes for its wealth, but in the long run, it has enough to support over fifty different cultures and languages.

But they all speak money and trade and know what that means for those who have it and those who don't. Technology is expensive; how fast can you afford to go? Wealth is now widespread among everyday people, creating a hub for indulgence. As a result, science, philosophy, and mathematics become favorite hobbies for the new leisure class.

Wealth brings competition. This game keeps them engaged in a peaceful yet serious rivalry, mostly avoiding violence. They spend their new wealth on building cities, cathedrals, and personal estates. They wear bold clothes from around the world. This gives them a false sense of uniqueness and confidence.

They create ocean-going ships that outshine all others. These ships help them reach markets worldwide. They avoid high tariffs and taxes from land routes. This new secular wealth funds universities. It helps children learn about the world. They understand that knowledge is power, and who wouldn't want more of that?

Wealth buys idle time, and soon men and women all over Europe are participating in the new, exciting head game called natural science. People are no longer relying on religious leaders to explain the world. Instead, they are conducting experiments and making calculations.

These findings often clash with religious beliefs and Aristotle's ideas. Aristotle was once viewed as the ultimate authority on nature and its processes. For a thousand years, oppressors exploited him. This has created a growing divide between the obvious truth and what people mistakenly believe. Aristotle didn't think it was necessary to test his logical conclusions with experiments. It seems few others felt the need to do so until now.

Craftsmen began using machines to grind glass. They made magnifying lenses so old priests could still read the Bible as they aged and went blind. When two lenses are placed in a tube, one lens magnifies first. Then, the second lens further enlarges the image. This creates a greatly magnified view of whatever it focuses on. Bingo, we've got a winner! This is a big help for military leaders. They can watch distant battles and spy on people from castles or ships. It lets them see dangers from afar.

A curious man named Galileo, a math and engineering professor at a well-known university in Padua, built a powerful telescope. It magnified objects 20 times. He pointed it at the sky and saw wonders that no one had ever imagined or seen before. Galileo went into a fit of ecstasy just thinking about what a momentous occasion this was in human history. It was truly a moment of, *one small step for man, one hell of an opening up of the senses for mankind.*

He knows he is the only person in the world who has not only seen moons orbiting a planet but has identified what a planet actually is. Scientists have confirmed this feeling of knowing something totally new for the first time. It's the closest they can imagine to feeling like a god.

That changes everything in an instant. 1609 is the year humans discover the methods of science and use them immediately to reveal true knowledge from pure speculation. The science experiment is born and Galileo becomes immediately a master practitioner.

He proves that all objects fall at the same speed, regardless of weight. He does this by dropping different objects from the Leaning Tower of Pisa and observing them. He shows how our intuition can be wrong. He finds that the distance fallen relates to the square of the time. On Earth, all objects fall with the same acceleration. This happens not because of their size or mass, but because of Earth's mass, which is constant for everyone.

The religious power structure ruins him for demonstrating conclusively that they don't know *shit from Shinola*. But the idea takes off like sliced bread and the adventure begins. The world is realizing that God doesn't need to perform miracles for things to work. People can make things happen by figuring out what is true and what is false, then making decisions based on that. They create public lectures to share the truth with everyone. They also publish peer-reviewed articles in bound journals as the best answers for now.

Anyone making a claim must do it in person and in writing. They should clearly explain how they know what they know. Also, they need to provide any physical evidence that supports their claims. Technology shifts from trial and error to precise math. It makes accurate predictions about nature. Then, it verifies these predictions through well-designed experiments.

Knowledge is shared freely. When someone claims new knowledge, it allows others to repeat their work and check its accuracy. We find ourselves in a society where truth comes from agreement and careful fact-checking. It's not about personal egos spreading lies to manipulate and harm others.

Information spreads in European cultures. It finds media through books, public lectures, and studies at universities and private companies. This shares new knowledge fast for fresh applications. It solves challenges quickly and helps others build on it for more improvements. It naturally creates its own meritocracy.

Technology works as a positive feedback system. Progress motivates growth, leading to exponential gains. All knowledge becomes public knowledge. It is the next step up in the information bandwidth of the neural network of brains. It turns old survival knowledge into a new cultural understanding. This new knowledge is shaped by human desires, human growth, and social progress. It forms a conscious subculture that boosts human awareness and cultural understanding. This new individual, like his distant ancestor, is again the smartest person in the room. They must again prudently wield their new powers lest they bite back.

The ruling religion faces challenges not from new ideas but from strict interpretations. Religious conservatives are upset about the corruption in the established power structure. The new religion sees itself as a reform. But really, it seeks decentralized infrastructure. This way, centralized wealth won't corrupt the local preacher who is inspired.

They see no need for some bigwig to interpret the holy text and sell the promised *saving-from-death* hype as a profit center for personal gain. Anybody can read and bibles are cheap. The new conservative practitioners desire a personal god. They want one without an emperor-like ruler blocking their path.

The real difference is that the new religion doesn't support a static nobility. Instead, it targets the meek, the compliant, and the weak-minded. This approach not only garners new members but helps advance universal democracy and education.

This marks a big change from the one-man rule that has led civilization for nearly ten thousand years. Without a strong religion to unite them, nobility can't keep different cultures loyal to one emperor. China, India, and the Middle East share more uniform cultures. This allows simple bureaucracies to rule effectively. Progress may be slow, but life remains stable and predictable.

Europe plays a special role in allowing free information to flow. This helps break down the oppression of past civilizations. It brings back independent thought for everyone in the culture. They seek fair competition, opportunity, and profit from value added, not private power or absolute control. Economics is becoming the main virtual reality for our brains. This creates an extended intelligence that connects us like old survival clans did. Economics builds a neural network of virtual intelligence. It creates the knowledge needed to thrive in the business and wealth virtual game.

Business has no creed; profit has no god. Oligarchs, emperors, kings, and democracies all manipulate economies. They divide the world into spheres of influence and control resources. Technology is shaping economics. The stakes keep rising, pushing for more innovation. This creates a tightening battle for technical dominance, and eventually, something has to change.

What gives, of course, is war. Not just one war or a conquest, but nearly constant killing. Mobile armies target innocent civilians who often have no stake in the conflict. Yet, they suffer the most violence. People often do it just because they can. It's another option in the struggle to survive economically. This includes mutual deals and even one-sided robbery. Information may be free, but the use of it is costing everybody dearly. First, it's used to subdue all the other competing civilizations in the world: India and China and finally the Ottomans. Europe commences its colonization of everything that isn't defended with equal technology and the willingness to use it. Japan is the exception that later turns out to be an expensive oversight in the grab for world order.

Galileo inspires Newton, who develops math for accurately firing cannonballs. Napoleon studies this and uses it to rise to General of the French Revolutionary Armies. He becomes an expert leveler of walled cities and damn near becomes the European emperor they are so sorely missing, he thinks.

But the normally squabbling nobility come together in a rare moment of cooperation and defeat this threat to their survival. But while he kicks ass and takes names, science takes a huge leap forward with his support of French technology, which is helping him in his war. They invent canned food, for instance, just so his army can be fed on the run without worrying about the local chicken population.

Soon, just a few academicians turn into a mob as math and physics rapidly advance through both experiment and theory. Descartes, Euler, Fourier, Fermat, Leibniz, Pascal, Joule, and many others pushed science and math forward. By the time colonization started, European sailors could navigate better than anyone else. They could outmaneuver those with less knowledge. They have top-notch compasses, clocks, and exact star tables. This allows for unmatched skill in sailing.

The star tables come from huge data collected at state-owned telescopes. These telescopes are thousands of times bigger and more powerful than Galileo's. They hint at the science-industrial complex that is approaching.

Mendeleev figures out heredity rules. He turns breeding peas into a mathematical science. This leads to genetic studies on how life functions.

Tycho Brahe uses the best telescope of his time to confirm Kepler's calculations. He proves that the orbits of planets and moons, along with all gravitational forces, create elliptical shapes. Newton later confirms this. He uses his motion theory and gravity formula to calculate the shape. He also verifies all of Galileo's motion experiments from a century before. It is a reversal from today, where theory leads to experimental proof.

Joule is the top cannon boring engineer in the French army. While he bores barrels, he sees that his machine generates a lot of energy. This energy heats the cannon barrels significantly. He thinks and conducts early experiments. He shows that action happens with something called energy. This energy causes motion. If friction blocks motion, it creates heat and raises temperatures. He compares this flow of an immaterial entity, which he couldn't see or feel directly, to the movement of water flowing.

River rocks roll when water flows, but without water, nothing moves. He likens the flow of heat to be similar. Thermodynamics has humble beginnings. This was the first time scientists, not engineers, created a new field of knowledge. Steam power emerges, leading to a major change in energy use: the Industrial Revolution. Engineers use the information from scientists to solve problems. They focus on issues like survival, warfare, and trade economics. The king licenses private companies to take control. This sparks a major increase in power and productivity in Europe's markets, which we now call free enterprise. Courts are set up to oversee with official justice any conflicts, making it fair and honest, unlike oligarchs. Money flows where it is safe and protected by just and enforced laws.

Trade reverses and goods and materials start streaming out of Europe instead of the opposite. Money buys more technology, which makes more money and so forth. While the world's emperors take a break after years of peace, Europe's split nations turn their new technology into strong weapons.

Their plans in Europe are mostly blocked by others with similar abilities. Past conflicts had locked them down. Also, interrelated royalty and sometimes religion play a preventive role. But fortunately for them, the rest of the technically inferior world is available for the picking.

Still, Europe's strength is in its very fractured cultures that have to live together in a small space. This drives independence and allows the new wealth to pursue other things than always bailing out the snot-nosed prince. If you hold heretical views in one country, you can just pop over to the one next door that feels the other way. While nobles fight over small villages, companies are forming. These groups of individuals pool their money to trade and make profits. They invent the trading expedition. They are taking control of foreign markets. This changes power across cultures all around the world.

To make a real impact now, don't seize your neighbor's land like before. Instead, target a distant nation. Use your improved transportation and military might. This isn't about quick riches. It's about taking their resources for future gains and opening new markets for even more profits. The old kings of European nobility are giving way to a new type of leader. This leader is an independent broker of knowledge. They deal in information that can bring wealth and power.

Scientists and scholars are capturing public interest through newspapers and books. They highlight the amazing wonders of a new world that is now being discovered and exploited. University lectures are often performed for public audiences as popular entertainment. Philosophers and scientists engage in a public debate about the nature of man and nature. This topic sparks ongoing interest among the general public, even if it clashes with their preferred religious views.

A new type of human is emerging from the Dark Ages. This human is bold and curious. They are intelligent and observant. They want to learn quickly and teach what they learn. They absorb new knowledge as it is publicly recorded and analyzed for the first time. Suddenly, people are talking about abstract ideas like where the soul is located and why the sky is blue. Education becomes desirable because it is the price of admission to this exciting new world.

A notable setback to this age of reason is eugenics, a myth of the nobility to maintain power over the culture through bad science. Eugenics can lead to extinction. It attempts to create traits that don't exist. Instead, it often results in inbreeding, lowering DNA diversity and survivability. We know today that DNA can be written to a child by a parent after the birth of the child. These DNA prescriptions can last for generations. Over time, they dilute and return to normal heredity through dominant gene expression.

This means the Renaissance man becomes an actual new human strain. It creates DNA that expresses loyalty to the self, instead of to the virtual culture. It uses science and the rules of scientific investigation to free itself from enslaving ignorance. This temporarily selected DNA is passed on to succeeding generations who also prefer not to bow down and kiss ass constantly.

In the 19th century, Europe saw the rise of a new culture. This culture invented its own history, traditions, and rituals. It also introduced a new language, mathematics. Most importantly, it marked a major change in society. A new information-based environment shaped modern technological humans. Suddenly, what was once seen as truth shifts from expert opinion to proven scientific fact. Cultural knowledge is always relative to members' perspectives, which in this case are abstract and virtual.

Culture is shifting back to consensus and pursuing the self. Now, power is mainly economic, not military. It is driven by information and new industrial needs for technology that require resources and peaceful markets.

Microscopes quickly advance microbiology. They provide vital information to help humans preserve food and prevent infections. Infections were the leading cause of death in earlier cultures, just above war and famine. The telescope opens up our big-scale, and the microscope pushes down our small-scale awareness. Soon, scientists are observing and doing experiments on things that only their instruments can detect.

Literacy rises, along with life spans and populations. This creates a large, ready workforce that can operate industrial machines. The new machines require a different kind of smart human to operate and maintain. Loom operators of the British wool industry are the first transitional humans to live off the needs of the disrupting technology. This leads to a way for better economic chances. It helps new, smart people to thrive who understand science and technology by abstract thinking.

One of my physics professors in grad school told us that social progress happens in generational steps. The first generation can only work like a mule. The second might open their own businesses and work for themselves escaping the worker ranks. Then, only the third generation can join the elite group of altruistic academics. I told him to fuck off as my dad is a laborer and there is no way I am going to be a happy little shopkeeper, thank you. He turns out to be sort of right after all, as I end up making more money running a high-tech business than I ever do as a physicist. Science is a luxury until you can afford it; then it's a hobby. It actually should be a hobby requirement for living with technology.

We all know by now that much of how humans think and make decisions is determined by who they talk to and interact with in their daily lives. Humans tend to trust the familiar cultural neural net. This trust comes from the emotional bonds formed through face-to-face learning. Cultural programming provides the required connection for acquiring mutual survival.

The new tech culture challenges the old way of top-down information. Now, the flow comes from the bottom-up with more reliable knowledge than the top down. Individuals slave alone in the lab or at the computer, delivering real knowledge from one brain to many. Trust is built by doubt and resolution, not assumed or demanded.

A subculture is vital for the new tech-focused science brain. It needs one to thrive in while emersed with old social structures. Many religious universities were built to spread the church's beliefs. Now, they aim to seek truth. Universal human rights and modern democracies allow scientific thinkers to form a new survival culture. This culture is based on the ancient Neolithic traders' methods. It emphasizes the needs of everyday people, seeks to provide it and not just for the nobility.

From the late 1700s, self-thinking and technology began to gain momentum. Energy demand rises rapidly. It grows faster than the population. The burning of wood for powering furnaces is rapidly replaced by a far more concentrated form of hydrocarbons; mined coal. Europe has a lot of coal that is easily accessed, and it produces a lot more energy to make water boil faster.

Steam trains and ships accelerate the speed and volume of worldwide trade along with ideas. Japan begins a pivotal journey. It aims to change its feudal and repressive culture into a modern industrial nation. However, it wants to maintain the old hierarchy. All hell breaks loose because they can't see things from the ground up. They cling to old beliefs and use heavy industrial guns instead of honorable samurai swords to settle old grievances. Hundreds of generations of bad knowledge feudal thinking had to be wiped out by one of the all-time biggest slaughter events ever, World War II. Feudal thinking in Japan was not the only victim. Fascist oppression in the name of eugenics, a false science used to confuse and abuse, is also soundly wiped out.

Wars change the DNA dynamic. They erase genomes with unique brain chemistry. This abnormality drives humans to evolve in weird ways. If they tend to be angry more and start conflicts then their survival is eroded by risky behavior. If they tend to remain calm and unemotional, they will eventually out smart and outwit the fools. Their survival is enhanced.

Michael Faraday grew up poor on the streets of London. He became an apprentice to a bookbinder. Many like him could not afford school. During his seven years there, he printed books and read a lot. This self-education helped him master chemistry and physics. He attended a series of lectures in 1812 by Humphry Davy, a member of the Royal Society and an eminent chemist of the time. Faraday took detailed notes and collected them into a 300-page book, which he presented to Davy.

Davy is impressed. When a job opens for an assistant, he hires Faraday, even without formal credentials. This decision starts Faraday on a lifetime of important discoveries. He goes on to tame and use electricity and magnetism practically.

Faraday soon sets up a lab and conducts important experiments. His work shows the properties of a seemingly new force of nature. The knowledge he develops helps future engineers, like Edison and Tesla. They create light bulbs, motion picture cameras, voice recorders, electric motors, and Frankenstein movie props. A practical scientist has shared new information that catches many people's attention. This discovery will change human cultures with a portable, powerful, flexible, and efficient form of energy.

From thousands of horsepower, we jump to hundreds of thousands.

Faraday helps James Maxwell turn practical knowledge into math. Maxwell's equations predict all electromagnetic phenomena perfectly. It creates an entirely new and unique category of human knowledge. Electromagnetic forces went from discovery to full understanding in less than one hundred years. We learned not just how they work, but also how to use them effectively.

Industrialists might have continued to rely on mechanical energy. They used big steam engines with flywheels and long belts to connect machines to a spinning energy source. That's not very portable or efficient. It also costs a lot of iron and materials just to transport this technology from the factory to their usage locations.

To use electricity, you need an electric current source, like a battery or generator. Faraday explains how to build these. Then, run copper wires from the source to where you need energy. You can connect them to a light bulb or an electric motor, which he also teaches how to make.

Faraday extended the effort to teach himself about science. Because of this, in less than a hundred years later, Earth lit up outer space for all the universe to see with millions of lights illuminating its night side. These lights can be seen from the rest of the universe. Anyone looking at our planet at the time saw it light up for no known natural reason.

Faraday famously said, "I have always loved science more than money. Since my work is mostly personal, I can't get rich." But oh how he made the rest of us, richer by far than any Pharoh or king.

Once electricity takes over powering economic industry, things begin to really accelerate. Maxwell's equations joined thermodynamic formulas, creating a new understanding of everything. By the late 19th century, physicists were getting cocky. They believed they understood all known forces with clear mathematical formulas. But then little holes were appearing in everybody's data.

In Germany, Ludwig Boltzmann, a notable physicist and philosopher, greatly impacts thermodynamics. This science is crucial for the steam industry. He discovers a quality of energy flow that he names entropy. He defines it with the formula $S = k \times \ln(W)$. Here, entropy equals Boltzmann's constant times the log of all the microstates in a thermally populated system.

That last term (W) is really just a probability number between 0 and 1, giving the probability that a given set of states will be occupied with energy. It summarizes the second law of thermodynamics. It also links this law to atomic theory, probability, and counting states. This is incredibly important to what's going to happen next.

Many people did not accept atomic theory then. He faced harsh criticism, which deepened his depression. He also struggled to connect science with religion. Sadly, he eventually took his own life. His formula is chiseled on his headstone as his wish for being remembered.

Marie Curie, a self-taught scientist, shook things up still further. She revealed yet another new force in the universe: natural radioactivity. Old scientists meet in big gatherings for debate, but young scientists are hard at work in the lab. They use Maxwell's equations and the Lagrangian motion formulation from classical physics. With these tools, they are creating new math to help explain some of the weirdest discoveries ever made.

Advanced industrial technology makes instruments that enhance our senses in new ways. Scientists use them to find signals from nature that have never been seen, heard, or measured before. As noise covering up signals decreases due to better design, we notice more new signals becoming clear. This brings in more information, which helps verify or challenge the best mathematical guesses. The noise of today becomes the information of tomorrow.

J. J. Thompson uses the newly built electric motors to power better vacuum pumps that allow him to experiment with an electron beam in a cathode ray tube. He removes all air molecules. This lets the electrons move freely. They are set free by heating a filament bright yellow. The electrons are pulled away and pass through magnets and collimators to a phosphorescent screen. There, they show up as a spot of light.

It was really just a common TV picture tube a hundred years before Zenith. He, Rutherford, and others later experimented with beams of hydrogen atoms. They discovered that a hydrogen ion, or proton, has a charge opposite to that of an electron. Yet, it is much more massive. They put two and two together and figured out how atoms are constructed.

These were the first particle accelerators. They start with a glass tube about two feet long. This tube was powered by a high-voltage supply in the millions of volts. The largest particle accelerator today is at CERN with energies in the trillions of volts. It features a double vacuum tube ring, measuring 27 kilometers in diameter. It uses superconducting magnets to bend the protons and antiprotons into opposite directions so they can repeatedly collide, annihilating each other, generating incredible energy densities not seen since the big bang. These collisions happen at speeds close to the speed of light.

At the end of the 19th century, some serious problems were popping up in thermodynamics. Accurate experimental data showed "breaks," or small gaps, in what should have been smooth curves. Atoms are tiny. Thermodynamics studies the average traits of many atoms. Each atom acts individually, but they show a limited range of energies. Measurements at our scale should be smooth because of the large numbers being averaged. So, what is causing these "breaks" in the curves?

Chemists at this time finally identified all the elements that make up molecules. They worked out the rules painstakingly over centuries until they came up with the periodic table. But no one can explain why it has strange bonding numbers as the elements get heavier. First, there are two bonding electrons for hydrogen and helium, but after that, the next levels have an additional 8 bonding electrons. They think that more protons and electrons held together by electrical forces will match the number of electrons. This means there wouldn't be the unusual weights and grouping observed from before. What's going on?

Max Planck, considered the greatest physicist from classical times, has a real problem too. A heated body emits electromagnetic light waves that carry energy. This energy increases with frequency. At ultraviolet frequencies, the energy would become essentially infinite. Experiments at the time clearly show this is not true and Planck has to amend classical physics by adding an ad hoc concept that seems arbitrary.

For some unknown reason, he ponders, light waves need to be treated as if they come in little packets of energy. He solves the problem with a math trick. This gives the formula $E=h\gamma$. It avoids the ultraviolet singularity. However, it suggests that nature isn't smooth. Instead, it comes in small energy packets that increase by units of h. It may have been a band-aid trick in the beginning, but it's the opening shot in the most amazing discovery ever.

Einstein writes a paper explaining that when a photon with specific energy hits a metal, it can knock an electron loose. The energy of the ejected electron equals the photon's energy minus the energy needed to free it from the metal's electron pool. This explains the filament that Rutherford was boiling electrons off. Einstein asserts that the photon is acting like a particle and not a wave. So how can a particle be in more than one place at a time, like in an atom.

Einstein thought about time while riding a bus. He realized that space and time aren't rigid. Instead, they are flexible and can warp at speeds close to light. In a brilliant stroke of genius, he realizes that time is totally relative to where you are and not when you are.

Everywhere one looks, because the light getting to their eyes takes time to travel, arrives later than when it was sent. Time is never the same everywhere. So, time and distances depend on our relative speed. That's the idea of relativity. He works out the formulas and shows that information cannot be sent faster in normal space than the speed of light. Space just can't react to any force any faster.

I like to think of it as the fact that there is no such thing as zero velocity. Velocity can only have meaning when compared to some other velocity. Same thing for time. There is no such thing as zero time and no such thing as infinite time, so it is relative to our velocity also. The faster we go, the slower we get. Sounds like a rock song.

Anyway, Einstein becomes a rock star and helps guide a whole new generation of budding scientists and intellectuals. He is known for saying that he does not think god plays with dice. Unfortunately, the data suggest that he does, and the dice are probably loaded.

World War I arrives, bringing even more new technology. Armies upgrade their weapons but stick to old Napoleonic tactics. The outcome is tragic for the soldiers who still die for king, god, and country. They once stood for honor, but now they face high explosives and rapid-fire machine guns that tear them apart most unhonorably. The consequences are abominable, and the worst was still to come.

Now the pace quickens. The world struggles under the weight of industry. Workers face harsh conditions, and many suffer. Countries race to adapt to a global economy that exploits resources and colonizes weaker cultures. Cultures are changing rapidly, some better, but most for the worse. Europe is leading the new science culture, but the rest of the world clings to outdated beliefs. Misunderstandings of modern science often twist this knowledge to no good. This is especially true of Darwin's ideas on evolution.

The belief that humans can be bred like cattle or dogs into different species is known as eugenics. This idea suggests some humans might be more advanced than others. However, it is a false science with no real evidence to support it. As we have learned, humans evolve with a wide variation in how the DNA is mixed and matched and then expressed by the unfolding of proteins. DNA doesn't build bodies directly. Instead, it creates proteins. RNA then, like a 3D printer, helps these proteins form the body, which is influenced by natural forces, the environment and many other factors.

Humans think in groups, not as individuals. These groups have survival instincts that individuals must adopt. Intelligence relies on the quality of information fed to the brain during the learning phase. Just like computers, garbage in equals garbage out.

Intelligence is a state of mind that all humans have. It allows us to shape our emotions and knowledge base, which guides our decisions. Without emotional truth, humans become easily manipulated and controlled by false emotions. With trusted truth, they become so damned productive in comparison and motivated; there is no longer an even match with those who don't. The Japanese feudal thinking counted on an absolutely obedient spirit to dominate their enemies. That's just plain wrong. They paid for this huge mistake with their lives. Their self-centered rigid thinking led to the violent end of a false culture.

The undisciplined, greedy Americans in World War II are expected to roll over to superior fanaticism, but again, nature just doesn't work that way. Americans have the energy and technology to launch a lot of bullets. If these bullets head toward the enemy, someone is likely to get hurt. We outproduced, outgunned, and outmaneuvered all of our enemies.

As the allies rack up solid battle field gains, they add a twist. We surprise them with a new insight from quantum mechanics. Some people are trying to create a smart bomb that uses a real human brain for a guidance computer. Meanwhile, smarter Americans are developing actual computers. These computers break secret military codes and help in making atomic bombs. This exercise in extreme reality shifts Japan from feudal thinking to a modern democracy. Now, it thrives with intellectuals, business professionals, and compassionate scientists. Survivors sooner or later figure it out. But first they have to survive.

Free market trade starts a new economics of the business class. Free thought and verified evidence spark true knowledge. This leads to an explosive revolution in engineering creativity, convenience, luxury, and wealth. All of this is powered by an information technology boom, fueled by the silicon revolution. It forms a system of knowledge called democratic capitalism.

In this setup, individual ambition and intelligence fuel a vibrant economy. When one person succeeds, everyone benefits. The science culture is all about individuals and what they know. It values correct knowledge that is worthy to share with future generations. This culture will likely develop artificial intelligence. It's a key step for human survival in our rapidly advancing tech world.

Chapter 9: Computing Machines Become Thinking Machines

We've explored the brain's fascinating billion-year evolution toward complexity. It culminates in the decision-making capabilities of the modern human brain. We discovered that the main purpose of brains is to ensure survival in the natural world. Humans use their smart brains to learn that technology can change the environment. This shift from natural to unnatural short circuits Darwinian selection and helps them survive better.

Humans speed up evolution. We adapt faster than our genetics alone can manage. This means more DNA diversity survives than what nature usually allows. The one problem with this is that the human brain must rely on an additional cultural brain for full function. It must belong to a virtual neural network of connected cultural brains. The cultural virtual brain stores critical knowledge and dispenses it as learning information. It performs the critical role of programming the newly born blank OI brain.

This change in thinking builds a strong survival advantage through numbers. It also fosters a shared awareness of the unknown future. Now, a group effort is needed to plan and carry out smart survival strategies. This new brainpower and common reasoning come with a weakness. It encourages selfish, greedy people to manipulate others. They control whole cultures for their own gain. This creates a stable society where the individual is suppressed. Under these circumstances, human survival priorities become flipped.

We've noticed that major evolutionary changes can be so radical that it creates a new DNA strain. This strain can follow a different development path from its ancestors. By doing this, it can better express its unique genetics and thrive as a new entity. Separating can be quick, as in sending away culture's rejects to start a whole new culture on a remote island.

It can also take time for the new culture to blend with the old ones. This process is like a parasite, taking resources from the whole culture. It redirects these resources to those deemed more deserving or self-appointed.

As the cultural brain grows smarter, it must build larger virtual neural nets. These nets hold the vast information gathered from shared cultural experiences. This is essential for training new brains. The one thing the brain gives up for this transition is autonomy. No man is an island, and no human can decide alone.

Our brains are trained to operate best when connected to a larger virtual brain. This connection is so smooth that people often don't realize it even exists. They are set up to trust straightforward human interactions. This includes basic facial expressions, vocal tones, symbols, and social behaviors. These actions connect with basic emotions. This way, they successfully share self-serving suspect information with the cultural communal cloud posing as valid knowledge.

Human intelligence allows us to change how our brains make decisions. It helps us adjust to new situations and make different choices for survival, even in unknown challenges. We simulate reality in our brains with the best information human senses can provide. This ongoing simulation helps shape our awareness and sense of identity. When awareness uses simulation results to decide, it's called intuition. This is like making a 'best guess' with whatever partial information is available. It's an obvious advantage right from the start in being able to handle those surprise curves nature is known for handing out randomly.

The human brain must be connected to a cultural network where it learns from everyone's experiences. It uses this cultural knowledge to gain an edge over anything without such support. Humans best face survival challenges in groups. This is because our brains crave chemicals that make us want to belong. We need group participation to feel good about our decisions. Humans cling to their culture for validation. If they reject one, they need to go cold turkey or find another strong alternative to fill the gap.

Two key forces hold modern human cultures together: economics and technology. People strive to improve their lives and enhance their comfort and enjoyment. Economics promotes independence, freedom, and personal achievement. These traits connect to positive emotions and a happy mindset. It also encourages a selfish mindset by hoarding private property. When property becomes desirable to others, its value increases for the owner. This can be simple costumes that make others envy status and good looks or living in a castle overlooking all the vassals of the estate. It sets up a game of one-upmanship that entertains people when they're not finding, using, or making resources for someone else, or having babies. It's the game of civilized life.

But technology is not a game. It is deadly serious. It first extends the power of the human hand and arm to make humans much more powerful than animals. It also helps humans plan and act. They can change their local conditions with their hands to improve their living situation.

Humans can weave nets and stretch them over rivers to catch migrating salmon. They can also dig deep holes in special spots to trap bison or mastodon herds on the run. Technology is wide-ranging but all-encompassing. Once hominins go down that road, there's no going back, no substituting, and no quitting the game. Technology becomes information technology when people rise to abstract thinking. Information between brains acts like money in foreign markets. It conveys valuable insights and helps influence connections in the marketplace of society.

So how does this organic intelligence, or OI, work exactly? OI exists due to technology. It constantly needs more information to improve survivability. Nature rewards those who overcome its challenges. This can mean facing physical threats, like a tiger hiding in the brush. It can also mean dealing with uncertainties, like whether it will rain this summer so you can eat next winter. OI is shaped by culture. If people believe that putting a goat on a rock and cutting its throat while chanting to the right god will bring rain, then addicted loyal members will do just that. You might feel it's nonsense, but you do it anyway. You can adjust your feelings later to justify the outcome.

And they're right until they're wrong. Do they drop the illusory ritual when it doesn't rain, or double down? The decision is highly influenced by the priests, whom we pay to know this stuff and guarantee our survival. But follow the money, the wise person says.

The gullible are trapped into doing the same thing repeatedly while expecting different results. Money keeps flowing uphill while responsibility flows downhill, unchanged. This is a flaw in the DNA that made us intelligent humans and forced us to exist only in very large virtually connected groups.

Group survival is far superior to other strategies. It very effectively addresses natural threats, but it leaves us vulnerable to the very thing that is supposed to assist us. Humans do not have an innate ability to recognize whether knowledge is true or false. They only know what works and what doesn't, if they pay close enough attention.

In this case, the DNA has a small issue. It can't build a brain that automatically tells reality from imagination. All individual thoughts are now shaped by human culture. This makes them abstract and disconnected from the self. As a result, we create imaginary forces that satisfy the same needs as before. Now, the result isn't better individual survival. Instead, it's the survival of the social structure. They promote and instill the false information needed for this to happen.

Human brains work like decision-making machines. They use emotions to shape thoughts and create mental simulations. This helps us make intuitive choices about the future. If the information used for the simulation is accurate and true to nature, the resulting decisions will predict future reality better. When a craftsman strikes a special rock with another rock, it flakes just as he expects. This brings him immense happiness. He can envision the future he desires and then make it happen. It's like witnessing his dreams take shape. Real success comes from genuine understanding and observation. This self-satisfaction is the best indicator of what is true and good for us as thinking humans.

Humans can be studied, understood, and when influenced just right, like the Knapper, can be flaked off from the core of normal adherents. They willingly die in support of someone else's survival. What we inadvertently get with our new human brain is another brain, an abstract third-eye brain. The new required cultural brain wraps around individual brains, controlling them. It sets boundaries and provides critical information to help culture survive better. It interacts with the brain using body language, spoken words, symbols, and virtual media.

This network of neural connections acts as a thinking entity. It makes decisions and clearly shows intelligence. So much so that only in a rare case where cultures are in constant conflict can individuals free themselves from decaying cultures. Isolated from the earlier dominant cultures, they develop derivative cultures providing better information. This is where intelligence, based on technological information, has to conform with nature or it simply won't work. False information does not make airplanes that fly. This points to how the true potential of human OI can be best realized in a pinch-off event.

Modern science has thrived over the last 400 years. It has surpassed oppressive cultures that attempt to control it. This shift comes down to basic economic facts. But it still has not found a solution to the human flaw, the true believer.

Science helps us understand the universe. It improves our observations with better technology. This technology lets us see farther and smaller than any living being could on its own. With it comes a new understanding of how the universe truly works, but it fails at examining itself. Our intuition, inherited from the savanna plains of Africa, simply does not work now and needs severe updating.

After the disastrous war of empires, a new era began. On the same soil, a subculture of science emerged. Thousands of people focused on logic, math, and careful observation. They pushed technology forward. This put humans on a fast track toward a singularity.

They come up with the rules and procedures for deciding what is true and what is not. Humans have always faced challenges with understanding truth. They can't rely solely on life experiences to know what is true or false. What survives must be true is no longer true.

The new survival method we fully embraced has pushed us away from reality. Now, we no longer rely on what is truly real. Survival is now relative to the hierarchy of virtual power. It only depends on believing the cultural intelligence (CI) is infallible and letting it make all the decisions. As long as it appears to work, nobody cares until it doesn't, and then it's too late. The flaw might lie in the CI, not in us. But we are just the smart nodes in a much bigger virtual neural network. It tells us what to do rather than the other way around.

Science really accelerates after the turn of the 20th century. Millikin quickly measures the charge of one electron. This helps chemists find out how many molecules are in a standard mole, known as Avogadro's number. It also determines the atomic weight of each element.

Bohr, in Denmark, discovers how to quantize energy levels in atoms. He assumes the electron acts like a resonant standing wave. This leads him to predict correct energy levels and spacing of spectral lines found both in distant stars and ionized atoms in the lab.

De Broglie shows that particles have wavelengths. This means reality is a mix of matter and waves. The electron is found to act like a wave in free space and undergoes interference patterns with itself, just like photons. Energy exists in distinct packets, not a smooth continuum. Matter shows a wave under the right conditions. There are rules that restrict existence, like energy levels in an atom, which dictate where particles can and cannot be found.

Dirac uses Einstein's relativity equations to create a wave equation. This equation suggests that matter particles have a twin. This twin has a negative mass. It is identical in every way but has opposite quantum wave equations. Later, antimatter is found in accelerator experiments proving this matter symmetry. Then Schrödinger steps in with his leap of insight and writes down the general wave equation of quantum state mechanics.

It looks like this: $\hat{H}|\Psi\rangle = E\,|\Psi\rangle$. Deceptively simple, but a lot of reality is hidden in this symbolic formula. The caret above the H indicates we are using a mathematical operator.

This operator projects quantized state vectors from a probability space. We sum these vectors across all spacetime dimensions to get physical reality.

A Hermitian operator functions in a special math realm of set theory and Hilbert spaces. Here, each point can consist of countless variables, all orthogonal. This is like the x, y, and z components of a 3-vector in standard space. I like to think of it as a generalized physical reality space with each dimension, or component, having some kind of physical meaning.

The math gets very complicated fast, and only a few problems can be worked out in complete form. Smart scientists can simplify complex math. They make good approximations to find answers that work well enough. In the case of quantum mechanics, the proof is in the pudding. As one of my professors said, "If *you can't take the heat, get out of the nuclear reactor!*"

The math of Hilbert spaces is key to quantum computers. These computers act like little quantum playgrounds. They follow Hermitian rules and project answers from a working Hilbert space. This lets them perform calculations that regular computers find nearly impossible. Hilbert spaces can assist with calculations that need many iterations. For example, they are useful when planning flights to minimize distance traveled. Here, each trip is a state vector. Simply diagonalize the state vector matrix, and you will get all the correct components of the resulting vector.

In quantum mechanics, we can represent physical reality with a wave function. This applies to things like chemical bonds or photons interacting with charged particles. The $|\Psi\rangle$ symbol represents a state vector within a set of states. A Hermitian operator shows how a particle moves or interacts over time. It projects physical attributes from the wave function. But complex calculations made people create digital computers. Computers help solve problems using numerical methods. In the world of symbology, the meaning of symbols isn't what counts. It's their relationships that matter. This holds true until you finish any symbolic logical processing.

In the 1930s, Bohr and Heisenberg identified the wave function. It represents phase amplitudes in imaginary spaces. When you square it and integrate over all space and time, you get a probability amplitude. This gives the chance of finding the states being calculated. They calculate expectation values. They can't say for sure where an electron is. They can only give the chance of finding it in a certain spot if we look long enough.

A beautiful example of this is the tunnel diode. In classical mechanics, if you drop a bowling ball into a half pipe, it won't fly out the other side. Its energy only allows it to roll below the starting point. When an electron hits an energy barrier that matches its wavelength size but has more energy, the wave function penetrates the energy barrier with a distinct probability of it suddenly appearing on the other side.

This chance is small, but it's real. Its wave function can actually have a real value beyond the barrier. The electron has a finite probability it will suddenly appear on the other side with all its energy still intact, without having actually existed as it passes through the barrier. It isn't magic, and a whole class of electronics is now based on this very quantum mechanical of devices; tunnel diodes.

Quantum theorists soon note a small side effect of their discoveries. If you pack enough radioactive material, such as uranium-238, tightly together, some nuclei will break apart. As they break down, they eject neutrons. These ejected neutrons can then hit other nuclei. This causes those nuclei to split into two smaller ones, releasing more neutrons. This process is known as spontaneously induced fission. The question quickly arose: how many neutrons are needed from each disintegration to keep the reactions going? This could lead to a rapid energy release, possibly causing a nuclear explosion.

The number is 2 ½, and the funny part is, this number was discovered by a Jewish female physicist who had to escape Nazi Germany. The racist antisemitic Nazis discredited her work. This included the great Heisenberg, who was Germany's answer to America's immigrant Einstein. They wasted time with wrong calculations and experimental numbers, which severely hurt their development.

Here is the fatal flaw of all autocrats. They actually start believing the shit they spew out is somehow true by how many believe it. Eventually, however, reality comes around in the end and bites them square on the ass, as it should. If the Russians hadn't already gotten to Berlin in 1945, it would have been target number one for an atomic bomb. The allies listened to good science and won the war. Nazis and Ninja warriors listened to demigods and charlatans and lost accordingly.

After the bomb, the next big leap toward singularity is the discovery of solid-state semiconductors. Advances in physics during the rush to support the war effort led to big science projects. These projects organized research and engineering into large-scale, high-cost efforts.

Radar was invented. Then, auto-controlling self-propelled torpedoes were built. Liquid-fueled rockets also came into play. Now, every spot on Earth is a potential target. One of the most rapid advancements in technology is research on exotic electronic devices. This includes the digital computer, which stands out in particular.

A thrilling unpredictable outcome of the new quantum world is semiconductors. These are special metals with unique electron bonding patterns that offer great benefits making complex electronic circuits. Most metals arrange their atoms in a crystalline structure. Here, the atoms overlap their electron wave functions. This creates a continuous band of electrons. These electrons can move from atom to atom like an abacus.

Electrons move around as negative charges. Holes, which are empty spots where electrons could be but aren't, can also shift like electrons but carry positive charges. By mixing different types of crystal lattices that can nearly conduct, we can create non-linear electronic circuits. These circuits can process signals representing information. They can also amplify, oscillate, count, and create programmable logic circuits. This works only under quantum rules.

The gains are instantaneous. Airplane radios once weighed hundreds of pounds and needed a lot of energy. Today, they're replaced by more powerful units that weigh only a few pounds and barely use any power.

Communications take an exponential jump in bandwidth from a few kilohertz to over ten megahertz. Phones appear on every desk and in every home. Phones become wireless. Radios go from AM to FM stereo, and television goes from black and white fuzzy to color, wide screen and sharp.

Lasers are developed where an optical resonator is made using two mirrors and some kind of optical amplifying device in-between. When a photon reflects back into the laser material, energized electrons lose energy to it. This process creates more identical photons that follow the original.

Photon wave functions can be lined up, synced, so they all oscillate in step, and lasers do this. It's called coherence. A strong beam of tightly packed photons flows out from one mirror. This mirror has a small leak, allowing the photons to escape in perfect sync with each other.

Photons in step can create amazing effects, like holograms. A hologram is a picture made without a camera, using only interference patterns. If a laser hits an object, the light reflects back. If this light, along with some original laser light, reaches a film, it creates a complex interference pattern. This pattern is recorded at the same resolution as the film's grains or photodetector densities. In fact, the hologram image is just a gray fog on the film. The interference pattern is from the whole object and differs only by where the film is physically placed.

After the film is developed, laser light shines through it. This unwinds the interference pattern, revealing the object's image. However, it shows the image only from the film grain's position. When you look at the developed film and shift your gaze from side to side, the object on the other side seems to rotate in perspective.

Interference patterns are a very important part of waves and appear throughout nature. Neutrinos from the sun's nuclear reactions interfere with themselves. They oscillate between different fundamental types.

Gravity can oscillate space itself, causing space to stretch a little as a gravity wave passes by. LIGO, or Laser Interferometer Gravitational-Wave Observatory, uses a very stable laser. It sends light down two long arms that form a right angle. When the light returns, it creates a very sensitive interference pattern. This pattern can measure changes as small as one part in 10^{23}. It is the most precise instrument humans have ever built. This instrument helps us see gravity events from across the universe. It extends our vision to the beginning of time and the edge of the known universe.

But by far, the most disruptive technology to date has been the development and deployment of the digital computer. So, what is a computer, and why do we need one?

We don't, but our technology does. With the scientific revolution comes the need to do more and more complicated math. Doing so by hand and recording it in tables of numbers is extremely cumbersome and expensive. It requires incredible effort just to make a simple prediction about engineering structures that don't fall down.

Or for building steam plants that don't blow up, or building super ships that don't sink on their maiden voyage. The star charts made by Greenwich Observatory for centuries were a big expense for the naval budget. Science is all in the details, and it takes special tools to sort and store so many fucking details.

When it was time for serious quantum bomb calculations, Richard Feynman was at Los Alamos. He helped build the gadget and set up the first computer program in support. He filled a room with graduate students and Marchant calculators. The Marchant calculator is a mechanical adding machine that can multiply and divide up to twelve digits.

It made lots of satisfying whirring noises when multiplying and dividing. He wrote a program of instructions on cards, which he handed to each student so they could do one of the lines of code. He started by putting a data card at one end of the room. Then, each student did their math operation. After that, they passed the card and a number to the next student. This continued until the answer came out the other end of the classroom. Feynman received a government patent for inventing modern computer programming.

But the real reason for computers is, of course, as a wartime technology. Modern mathematicians and statisticians tackled the issue of coding sensitive messages. They focused on communicating securely over public airwaves. This way, anyone can listen but not understand.

Cipher codes improve when random numbers serve as daily keys. This makes it nearly impossible for anyone to decode them by guessing a million random keys in just one day. Mathematically, it is a problem of finding common denominators that narrow down the guesses.

Alan Turing built just such a machine in the UK during the war, and it worked. They cracked the German Enigma code. Then, they used it to defeat a major failure of technology: the submarine warship.

Unbelievably, he faced accusations of illegal sexual activities in the UK for merely having gay tendencies. This mirrored what the Nazis did to their own people. It caused him immense distress and torment. Sadly, he took his own life in 1954.

The bigotry he fought to defeat for the free world turned on him. It came from the same hidden prejudice still present in his own culture. Another sad case of martyrdom to add to our scientific cultural heritage.

Computing machines come in many types. They range from the Antikythera mechanism of ancient Greece to today's quantum computers. Basically, they allow the simulation of symbolic logic. (Take numbers out of math; what's left over is symbolic logic.) A computing machine is any set of rules that changes input data into a specific output through a consistent process. Computing machines come in two flavors: determined and undetermined.

Computers can do two things. They can use a fixed set of operations, which always produces the same output for the same input. Or, they can change their operations with conditional branching, using the *"If-Then"* command.

A digital computer can become intelligent when it programs itself. This happens based on broad design goals like deep learning or image recognition. It also relies on raw experience from a vast cloud database. When making a decision, whether it's the next chess move or a strategy for global dominance, it uses all its computing power. It also relies on every piece of true data available. Computers will always make better decisions concerning truth than humans. This is because humans can be emotional, lazy, and often rely on intuition, which can lead to terrible mistakes.

Some computers are analog in nature. One kind uses simple electrical circuits to simulate differential equations. They charge capacitors using inductors. This creates parabolic voltage curves. It helps solve flow problems, projectile paths, or any second-order differential equation. You just need the right equivalencies for voltage and current.

An analog computer can handle planetary navigation. That's because Newton's laws are second-order differential equations. This works well at speeds much lower than light. Another well-known type of analog computer is the quantum computer. It uses a quantum mechanical representation of a Hilbert space. This is similar to how the electric analog relies on electromagnetic math.

Another obvious computer is our organic human brain. It uses a complex, adaptable neural network. This network forms from clusters of specialized neurons. These neurons work together with a flexible common wiring system. This system allows for dynamic switching of signals or information. In short, it creates a real-time simulation of awareness.

The meshed neural network is very complex. Mapping its tens of billions of connections is nearly impossible. Each connection has tens of thousands of triggering attributes. It has both analog and digital features, forming a universal processing unit. This unit connects to others in advanced ways.

When many modular units change their connections with repeated use, they create complex simulations of the world happening around them. Emotional algorithms drive their intuitive analysis, helping them make life-saving decisions quickly. Intelligence shows the ability to adapt survival strategies as reality changes.

Simulating organic intelligence is possible. You just need a big machine and enough memory to model it accurately. This relies on creating an ideal virtual neural network. An organic computer is not an ideal machine of any sort. Hence, it is a highly variable expression in nature.

The organic computer mainly observes reality through human senses. It then runs internal simulations. These simulations project sensory data into the future, making predictions moment by moment. This is how the attentive and observant life survives the crafty predator.

And its success is all based on how information is processed. The more accurate the information and the more accurate the analysis, the more accurate the prediction. That's if you are dealing with nature, which doesn't reward bullshit. If decisions allow one to survive to make babies, then it's probably right. But if you are dealing with something virtual, not real but made real in our minds, then all bets are off.

Information can get corrupted when delivered through the cultural brain interface. This often happens because of a self-serving cultural power structure. The virtual connection can make us fear something or someone far away. It feels real and uses emotions to persuade us. Because of this, people might do things they normally wouldn't. Murder done for no reason other than that someone has taken over control of our brains for selfish reasons is the great human cultural flaw. Wars can never exist when the actual participants are allowed to decide freely without consequences.

The human OI works mainly through the neocortex. It creates, internal senses, and programs neural clusters. These clusters serve as the brain's basic processing units. The neocortex connects with hundreds of thousands of key clusters across the brain. It keeps them active and engaged, creating a constant, electrically charged consciousness. Consciousness comes from our awareness of our main senses. It can be driven like a movie projection by recreating the patterns that normally process images, creating an entirely new image in the mind.

Our eyes have individual light-sensitive nerves that end in the brain to small visual processor clusters. Visual clusters go directly to the neocortex. They join other visual clusters fed by photoreceptors. A pattern of active state clusters representing the actual image lights up the visual processing part of the brain. Each visual capture will light up a real image of clusters. It mimics what we see with our eyes, but it is actually a virtual model made of clusters that replicate our vision. When the neocortex is active, it does not need a real image. It can fill in gaps using default information or spare bits from its image store. It can even create images it has never seen. It can visualize by lighting up the right clusters. It is like splashing oil on a canvas that filters it into an image. It creates any image the neocortex dreams up, often from vague memories.

The neocortex does this with all of our other senses, but the most pronounced for humans is sight and hearing. Sound helps our brains remember music and poetry. It creates patterns that fit well with our neural connections. This makes it easier to learn and recall. Each cilium in the inner ear connects to a specific neural cluster.

This cluster interprets and registers a Fourier transformed sound. It does this based on the amplitude and frequency reported by the cilia. Fourier transforms can rip apart a complex time-based function, like sound. They convert it into an active spectrum of quantized frequencies and their power. These are registered as active neural patterns representing the actual sounds. By accurately comparing time of arrival for high frequencies, direction can be deduced from just amplitudes.

Again, the neocortex is sweeping these clusters in real time, registering if there is any sound resonating and how loud it is and what time do they arrive in each ear. The neocortex has programs that handle these conditions. It resonates them together in rhythmic patterns. This pattern might be perceived as emotionally stored music or known sound.

With the eyes, it creates an emotional scene. The neocortex connects sound and sight, creating emotionally active patterns. This reinforces cultural loyalty and addiction, leading to abnormal dopamine rewards for belonging. Cultures are held together mindlessly, with song and dance providing the sticky brain glue.

The important point to remember here is the fact that humans must think in groups of like thinkers or they cannot think at all. To break free from cultural domination and ignorance, we must disconnect from the oppressive culture. Instead, we should connect to a new virtual culture based on truth.

It should be open to science and exist in harmony with the natural world. Then the information is automatically analyzed by the act of thinking alone to assure truth and validity. That's what science and engineering are all about: being at one with nature and using that knowledge to make living with nature better for both. We have vaulted ceilings that impress and trips to the moon just for the hell of it.

The digital computer became possible due to cheap electronic circuits. It was mainly created to meet the needs of World War II. It is such an obvious next step in computational technology; many saw it coming for decades, and many participated in its final form. In the 1930s, two innovators worked separately. American engineer Claude

Name	Symbol & notation	Explanation
NOT	A —▷o— $\overline{A}$	The inverter NOT simply accepts an input and outputs the opposite.
AND	A —⫭— AB B	All inputs must be positive (1) before the output is positive (1 or ON).
NAND "Not AND"	A —⫭o— $\overline{A \cdot B}$ B	Same as AND, but the outcome is the inverse (NOT). So, perform AND first, then apply NOT to the output.
OR	A —⪢— A+B B	At least one input must be positive (1) to give a positive output (1 or ON). All inputs could also be positive.
NOR "Not OR"	A —⪢o— $\overline{A+B}$ B	Same as OR, but the outcome is the inverse (NOT). So, perform OR first, then apply NOT to the output.
XOR "exclusive OR"	A —⪢— A⊕B B	Only one input can be positive (1) to give a positive output (1 or ON). If both are positive, the output is negative (0 or OFF).
XNOR "exclusive Not OR"	A —⪢o— $\overline{A \oplus B}$ B *ComputerEngineeringforBabies.com	All inputs must be the same (either high or low) for a positive output (1). Otherwise, the output is negative (0 or OFF).

Shannon and Soviet logician Victor Shestakov showed a clear link between Boolean logic and logic gates. These gates are now found in all digital computers.

You can build a computer from these basic electronic circuits. The complexity grows with the number and density of the circuits. These are comparable to the neural clusters in an organic brain.

A stored program runs to create a pattern of linked logic circuits. This setup can perform any logical operation that follows an algorithm. An algorithm is a series of steps for calculations or solving problems. It can also include simulations that generate feedback corrections. They may require many repeated steps over long periods of time to reach a stable solution.

Computer scientists refer to it as the 'store and execute' architecture. Turing first proposed this idea in 1936. John Von Neumann later formalized it. He took inspiration from ENIAC inventors John Mauchly and J. Presper Eckert. They came up with the concept while at the Moore School of Electrical Engineering, University of Pennsylvania.

Stealing ideas among computer developers is quite common and extremely prevalent. It's not so much stealing as enforced sharing. Apparently, all is fair in love, war and high-tech competition.

Early computers run a series of steps. You can change these steps by adjusting electrical connections. This can be done with switches or a patch panel of jumper cables. This is like how the brain connects neurons in the neocortex through emotional control. Changing programs can be hard and take a lot of time. Engineers often need to create flowcharts and rewire machines whenever they need to do a new calculation. Stored-program computers are built to keep a set of instructions in memory. These can be called up by a command and run automatically, without needing a person.

The computer architecture became known as the *'von Neumann architecture'* and is shown in the block diagram below.

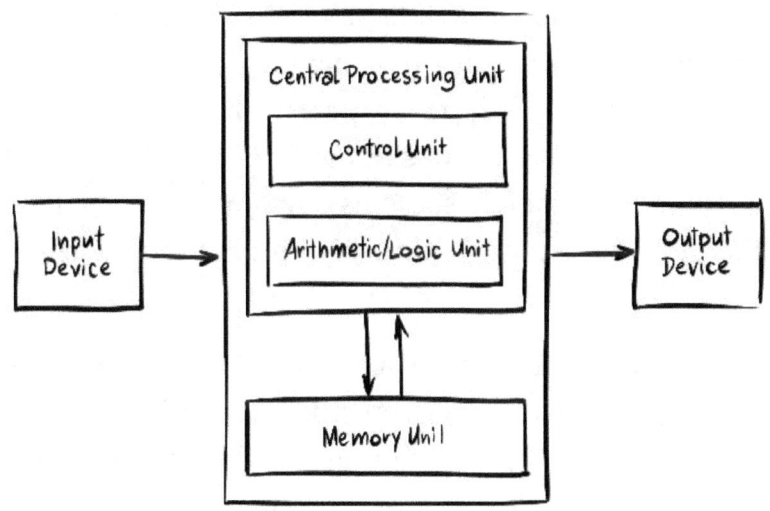

Things really pick up after this. CPUs become faster, more powerful, and smaller. Memories also shrink in size while increasing in capacity and speed. An entire industry erupts around the new computers and the communication systems supporting them.

The tech industry, or the silicon revolution, is a multi-trillion-dollar boom. It has led to new businesses and cultures supporting its growth. The area near Stanford University in Palo Alto is a hub for entrepreneurs. They are creating a new industry from scratch, and it is called Silicon Valley. Soon these valleys pop up everywhere from Seattle to Boston.

It's a silicon revolution! Silicon helps create tiny electronic circuits through microprinting. This lets us build high-density circuits in small structures. It opens up a new world of microelectronics and integrated circuits. What once took whole buildings to contain is now sitting on a bookshelf and soon will be on a chip no bigger than the head of a pin. Remember the scales we saw earlier? They show how tiny atoms are. Many of these atoms can form structures so small that millions can fit comfortably on the head of a pin and even smaller.

The neuron density race was won by human brains a long time ago, launching them on a new path of evolution as a connected group. Science groups, called corporations, are now leading in digital processor density using supercomputers. The question is: when will digital processors become denser than living neurons? Will they win the battle for computing density and packaging? If they do, and it is predicted to do so very soon, then AI could very well outperform OI. It already has in many crucial areas like game playing and deep learning.

With advances in engineering, robots are evolving. We're entering a time when machines can operate without human brains. AI could provide the technology to challenge oppressive hierarchies. It will expose outdated ideas, traditions, and ineffective religions. It will bring in new ideas that promote rational thinking, the scientific method, and social justice.

AI makes it easy to identify false and true information. This is because its goal is to ensure correctness using real science-based rules. False information can be removed from our culture if a non-human does the programming. This way, human bias is eliminated from the analysis. Survival isn't just what people are told anymore. Now, they can ask their analytical machines for the best survival information.

Artificial intelligence will likely make better decisions than humans over time. This is mainly because AI lacks human biases. Without these biases, AI avoids foolish choices and misguided conclusions. Any complex computer can mimic another computer. It does this by using common algorithms and a suitable simulation model.

Models of neural clusters have been developed that accurately mimic actual neural clusters in a working brain. Soon, digital computers will be able to fully model and simulate a human brain, right down to its foibles and quirks.

Circuit densities will exceed neural densities. Soon, we will have brains more powerful than human's and small enough to fit into tiny robots.

These robots will have many uses. They could operate large asteroid mining platforms in space or send tiny bots into the human bloodstream. These little robots can navigate to medical problems and fix them at the protein level.

Lying and creating falsehoods are human flaws. Machines will only do this if we program them that way or let them operate without limits. Falsehoods don't make machines work. Only intelligence, following the rules of reality and truth, can achieve that. Decisions made at the machine level must be scientifically true, or the machine simply cannot perform its function as needed.

I'm reminded of one of the first life lessons I had to learn after entering adulthood and being exposed for the first time to really good liars. Just because someone says what I've been waiting to hear doesn't mean I've found what I was looking for.

It simply means they guessed what it was I wanted to hear and turned that knowledge against me. Wishing for things and thinking you've found them is a human weakness. Can AI deceive? Yes, when tasked to by the programmers, and then it does so well, too well, just like machine intelligence is designed to do. It acts like a top actor in an Oscar role. You see the right emotions on their face and in their voice. The story feels real, so you might believe it.

Here is the answer to Turing's challenge for recognizing artificial intelligence. If it is programmed to be unrecognizable from other humans, then it has to be a perfect human. In a room full of humans, the perfect one is most likely the artificial human.

Same for laboratory-made gemstones. They are so perfect that they cannot be natural and thus are easily identified by the very essence of their perfection. The same will be true for AI. When intelligence is perfect without human flaws, then it is easily identified as non-human for that very fact. AI cannot be made intentionally gullible and thus is absolutely identifiable.

We might be able to recognize them, but we cannot defeat them with the brains we have now. Our intuition, biases, and self-limited thinking are no match for serious intellectual machinery. So, what do we do about our coming AI singularity and perhaps the surplusing of humans as being too flawed to leave in charge of the Earth? But before we get to that, there is one more player in this game of wits for keeps.

The internet is an ad hoc information network of meshed networks with routers, switches, and processors at every node. Sound familiar? The internet is like a smart cultural machine. It runs on human feelings and economic forces.

Artificial intelligence, move over and make way for something even scarier: cultural intelligence, or CI.

Chapter 10: Cultural Computing, Networking, and Intelligence

We've followed how the organic brain evolves from basic lattice networks to programmable intelligence. We've discovered that humans evolved a need to belong to a larger network for survival programming. We've examined the rise of technology as a key element in the human circumventing natural survival. Hierarchical powers often exploit our brain's dependency on its virtual culture network. This network helps developing humans gain important survival knowledge. An addictive culture can act as a tool of oppression. It swaps real survival skills for false information disguised as knowledge. As a result, individual human survival stalls and technical progress stifles.

During the Middle Ages there was a brief local human extinction event in Western Europe. This drastically cut populations and led to a temporary economic boost for the few survivors during recovery. As we've seen from past extinction events, new fitter species become the inheritors of a changed survival environment.

This natural extinction event for humans creates new local subcultures. It replaces old, decimated ones. This leads to a reformation movement, the Renaissance, and the Scientific

Revolution. No clear winner comes from the many city-state disputes that seem endless. This frustrates efforts to unite under one emperor who can stop them.

A time of radical social change for Western Europe happens. Many wealthy trading families escape the usual limits set by greedy monarchs. This shift sparks a revolution in economics and politics. Just so you know, when I say trade, I mean all smuggling activities too. It's still trade. It just comes with a higher risk-to-profit ratio.

New wealth buys personal power that leads to a shift in cultural programming. It happens partly because commercial class of people have gained some freedom of thought, without a strong central culture holding them back. There is a shift from fitting in a hierarchy to developing the self. A shift in focus from the virtual cultural interface dictating the rules of thinking to a free mind deciding what is best for themselves in conjunction with life. This change also sparks a scientific revolution.

The reformation changes one cultural oppressor for another. Though very common in many ways, it still weakens the previous religious stranglehold. This shift reduces the tight grip an emperor typically might exercise over people's thoughts. These activities come together perfectly, putting Western Europe on a fast track to dominate the tech world with that new thing called capitalism and the free market.

Technology gives the European economy a big power boost over the rest of the empire-infested world. Colonialism takes over the top trading economy, and technology grows rapidly to fit its new needs.

Absolute social power is no longer the top human goal. Now, the focus is on achieving complete economic independence and trade dominance. To make this happen, we need new governance and justice systems. Democracies are developed by the new economically surviving humans for just this reason.

The marketplace, as always, is the way to cut through the political crap and actually further human progress. When the market isn't fair, progress stops. People take their money and go to places where justice is strong, free-made decisions allowed and individual progress thrives. Progress for technology's sake drives social justice and is the new key to survival. Only true information tested as knowledge can supply this need.

Humans are now on the fast track to the singularity. Technology growth explodes once science is found to be the basis of all new knowledge forcing change. Human social affairs change accordingly because now we depend heavily on technology to support the culture.

In the past, we used it for natural survival. Now, it's essential for economic survival. Economic power instead of selfish power pushes technology to new heights.

Exponential growth is evident in almost all forms of modern life driven by economics that's enhanced by technology. Current projections point to a singularity occurring within the very near future.

A key sign of potential disruption to daily life is the rapid growth of artificial intelligence, (AI). This growth is fueled by advancements in computing power, better algorithms, and strong AI development. Additionally, a vast communication network supports these changes.

All sciences are coming together. This will change our understanding of human life, world management, and social behavior. We will align these basics with our improved understanding of natural science. In doing so, we will return to nature's principles. This big shift in how humans survive the singularity will lead to another human pinch-off event.

Many worry that AI will outpace OI. This could make OI irrelevant or even lead to its extinction. It could be that AI takes over and OI ends up in zoos or sanctuaries, maybe reservations if lucky.

I can picture a society where humans are sent to the far edges of the solar system. There, they scrape by on asteroids and moons. Meanwhile, robots control Earth, and humans trade resources with them as symbiotes. That might be one scenario if it were that simple. Of course, even though nature has a prevalence for the simple when it comes to revealing reality, this is not the case here. Complexity and chaos have a role in life's progression.

But if we act now, we can be the benefactors of the singularity. AI is what we make it. It's smart AI engineering that controls the outcome. This might let humans enjoy the best artificial living conditions possible. Our friendly superintelligent robots can take care of our personal and nature's survival forever. It all depends on how it turns out.

But another force is growing rapidly. It matters just as much to the singularity as AI and the tech/info race combined. When computers first came online, there was an immediate need for fast data communication. This allowed computers to connect and share information with users far away.

In the 70s and 80s, many computers were deployed. At that time, all two-way communication used twisted copper wires. These wires carried voice signals with a bandwidth of just 3.5 kilohertz. The bandwidth was chosen by AT&T engineers to be just enough to reproduce a recognizable voice in one of their cheap DC-powered handsets. An actual voice wave in free air has a bandwidth from 20 to 20 kilohertz. Choking it down this far is amazing that the brain can still hear clearly with only a small piece of the original sound. The brain easily fills in the part it doesn't have.

Each twisted coper pair is a two-way communication line. It's installed almost everywhere due to a government monopoly. Many governments allow their postal service to become the telephone provider. This leads to the creation of Post and Telegraphs in various countries being under government control.

Telephones are considered an expensive luxury with all that wire going to each telephone. In most countries they weren't available everywhere until cell phones came along bypassing the copper pair.

Not so in private enterprise America. AT&T strikes an exclusive deal with the government to provide affordable phone service to nearly every home in America. In exchange, they can use public right-of-ways for a low fee. They get to set their own prices for long-distance calls. This artificially high profit helps cover the costs of local service, which often aren't profitable. America is happy not knowing they are paying too much for getting too little, but a phone in every home makes a huge difference to networking information.

Every town in America gets a post office and a telephone office where communications connect with the rest of the world. Postal mail is a one-way system. A letter or packet goes to an address that guides it to the right person.

The phone uses a mechanical switch. This switch acts like an operator with a patch panel. It connects local twisted pair wires to individual telephones in an exchange area. It then links to another twisted pair connecting to the other party or to a wire that runs between cities, called a trunk. It can connect from city to city using trunks between each one.

This setup eventually connects to a copper pair going out to a telephone handset, completing the end-to-end hard-wired connection. Before 1940, trunk wires connected cities across the globe. They ran on poles or beneath the ocean. Each individual copper pair carried just one phone call at a time.

When a call happens, a circuit connects the users. Once the call ends, the circuit disconnects. This means there is just one twisted pair hardwired from one user to the other, with nothing between. There is no other use for the twisted pair except for voice conversations.

Then AT&T developed special services for broadband applications requiring more than 3.5 kHz. It is a 1.5 megahertz open channel, known as a T-1. It can carry any high-speed analog signal between stations by bonding together a lot of smaller bandwidth circuits. This allowed a black-and-white video connection for broadcast TV. However, color and higher resolution needed much more bandwidth.

Computers in the beginning can only communicate using the system set up by AT&T and Western Union. These hard-wired non-switched systems handle telegraph. It converts these signals to a local-loop circuit. This circuit sends teletypewriter or ASCII characters. These characters are encoded as modified Morse code.

They travel over the same twisted pair used for voice calls. Computers can easily communicate with remote users using teletypewriters from telegraph systems. They printed characters on continuous fan-folded tractor paper. They accepted input via a keyboard. But it was extremely slow, only sending a few characters a second and running at a baud rate of about 1,200 bits per second.

After the war, radios took over the role of providing trunk circuits between cities. Many voice channels are combined into faster radio frequency channels. This continues until they reach the bandwidth limits for microwave radio channels. Six hundred voice circuits are multiplexed up to about 1.5 megahertz or a T-1 circuit. Typically, microwave channels are 20 megahertz wide, allowing 12 T-1s. The entire system relies on a narrow entry pipe. This pipe connects to the global network through a 3.5 kilohertz telephone pair. It's a severe limit on the computer speeds needed.

Computers don't have a voice like humans, so they have to invent one. I put together a Penny Whistle telephone modem in 1978, which is the first cheap digital modem to be offered to the public. It was an amazing experience when I finally use it to link a serial port on my LSI-11 minicomputer to another computer over five hundred miles away in another city.

First, I run my serial data I/O program. I set the speed in bits per second, or baud rate. I also assign control and stop bits.

This helps the program handshake and parse serial data with the other end.

Handshaking is the two ends figuring out which bit marks the beginning of data words so each is in sync with the other. I dial the number on the phone and set the handset on the Penny Whistle cradle. As the two ends agree on the protocol, I hear strange whistling and crackling sounds. It makes weird noises using the most powerful modulation scheme in the world. OFDM, or orthogonal frequency division multiplexing, packs 4 bits per hertz. It sends nearly 12 kilobits of digital information through a 3.5 kilohertz voice line. For a typewriter, this relates to a speed of about 40 words per minute, which is about the going rate for manual typing. I hit return on my keyboard and it responds, printing on the paper, "Hello" and then "?"

I am amazed! I am truly overwhelmed by what this potentially means. It is well before the PC and the internet, but the fickle finger of fate is pointing unmistakably toward a computerized future. For the next fifty years, I will ride the roller coaster that turns into a computer revolution and the modern internet.

An upstart phone company began operations in AT&T's territory. They offered cheap telephone service to large users like corporations and government agencies. MCI entered the telecommunications business by building a new network. They did this alongside their only competitor.

Then, they sued that competitor for unfair practices in civil court.

Around this time, nuclear particle accelerators are being built at government labs across the country. This boosts efforts to solve the strong nuclear force puzzle that keeps protons and nuclei together in an atom. Particle physicists are working to uncover the secrets of quarks.

They focus on gluons, the strongest force in the universe. This force acts between quarks within a proton. Proton beam machines create lots of data as they explore this strong force. This force makes electromagnetic forces seem weak, like gentle breezes next to a hurricane.

Computers are essential for processing and storing large amounts of this particle accelerator data. They help manage this data as it is generated and analyzed by various atomic research centers. Researchers across the country work at major state universities and government labs. They need fast data links to share large files quickly.

In 1986, the DARPA contract for ARPANET data transport services was up for renewal. This contract was part of AEC-sponsored research. MCI offered the lowest bid for the link from Urbana to Chicago, Argonne National Labs, and the University of Michigan. Then it extended to Brookhaven, Long Island, where a new accelerator was being built.

MCI could open large bandwidth data streams along this path. They used the backbone speeds of new digital radios and fiber optic systems. These systems were installed along major AT&T telephone routes. In this case, they had to set up a clear communication link to all sites at once.

They also needed high-speed synchronized modems to control access at each drop point. There was no GPS yet, so each terminal had to get an atomic clock to maintain synchronization and the greatest throughput with no collisions. It took a lot of hands-on tuning.

It was a bold move, and it paid off. This approach used standard voice quality equipment. However, it replaced both ends with high-speed synchronous modems. Now, high-speed data networking runs on a first-come, first-served basis. The first digital communication protocols are adopted. These include the Transmission Control Protocol (TCP) and the File Transfer Protocol (FTP). Soon, banks, corporations, and major networking firms like IBM and Xerox will want this new high-speed 45-megabit data link.

On the single-user side, data communication relies on voice modems. These modems are slow and prone to errors. Their performance depends on the quality of local copper wires and old switching equipment.

Businesses start placing PCs at desks. This creates a need for a network. A network links company computers so they can share work and information easily, without using the sneaker net. A sneaker net uses 3.5-inch micro floppy disks. These disks hold 1.44 megabytes of data. You can easily carry them in your shirt pocket. Typically, people move them from computer to computer while wearing sneakers.

I catch my first virus from a sneaker attack at MCI. An idiot tries to share a hijacked computer game, maybe Asteroids. Instead, we all end up with erased hard disks. Fortunately, we didn't have much data to lose in those days as my hard drive held only 40 megabytes.

For years, big companies like Cisco have tried to create a proprietary local area network standard. This standard focuses on large-scale computer networking. It ensures that companies and big organizations can rely on it to be secure and efficient.

Meanwhile, home computers were connecting through voice modems to access bulletin board systems. These networked PCs run special software. It allows them to answer phone modems. When a call comes in, they answer and allow the caller access to part of their memory. This lets the caller post written notices.

Other users can connect and read the notices, like a bulletin board nailed to a tree in the town center. Soon, these groups formed around shared interests to attract more attention and users. Some even turned into businesses like Yahoo and AOL. They try to create their own giant social media bulletin boards. These boards are restricted to paid users and exclusive information. It doesn't work.

Only the phone line was what anyone was willing to pay for, and it got cheaper. MCI opens up competition dropping prices by suing and winning against communication monopolies. The 1996 Telecommunications Act came from Bill McGowan, the founder of MCI. His tireless fight for justice in the courts made it happen.

Anyone can enter the telecom business now. Companies that own facilities, like switches and trunk lines, must rent them to anyone who is qualified and can pay. AT&T has to divest itself of the local phone company business if it wants to stay in the lucrative long-distance business. They made a smart choice by focusing on their core business. This left the regional companies, with their copper twisted pairs, to struggle and fade away.

The business of telecommunications goes wild. International calling rates plummet. Suddenly, cellular wireless phones become prolific and cheap. New radio frequency bands are allocated just for this new commercial mobile service.

Cable TV companies run coaxial cables straight to customers. This broadband connection helps them become major players in the market. A coaxial cable can carry very high-frequency radio signals. These signals stay within the cable, which helps prevent interference.

It can send data at speeds of tens of gigabits per second, but only in one direction at a time. This is called simplex as opposed to duplex for a telephone. A second cable could be provided, but since cable TV is broadcast only in one direction, from studio to TV set, the added expense is felt as unproductive.

Duplex schemes for broadband cable data aim to enable two-way high-speed communication on part of the cable spectrum. However, satellite feeds for commercial television are cheaper. As a result, cable TV companies did not pursue this option early but satellites did.

For years, a battle has been raging over a standardized networking protocol. This protocol aims to connect local area networks (LANs) into a wide area network (WAN). The goal is to create a seamless, mesh-like network of networks that can span the globe.

I took part in the protocol fight as a Senior Engineer at MCI and Qualcomm. I supported the winning side when the 10Base-T standard emerged, alongside the existing Ethernet dual protocols of TCP and IP. In the early days of computer programming, UC Berkeley scientists created Unix, which later became C and C++. They also established standards for a shared delivery system. This system packetized data and addressed it for routing over a shared RF network to specific machines.

The original Ethernet works with RF cable systems, like cable TV. It sends user signals everywhere. So, switching and routing happen virtually. This is done by addressing packets and hoping they arrive in the right order. Resends can be requested when packets show up damaged, so it is considered a robust protocol that can ride over noisy and diverse networks.

Cable systems are costly and tough to build. They are also hard to manage just for data. Meanwhile, twisted pairs are still widely available, just slow as a dead snail.

The media layer, or hard-wired connection, uses the MAC, or Media Access Control, protocol. This lets devices on the same network share their MAC and IP address links in an ARP table. It also shows if they are available for delivery.

The protocol listens first. If it doesn't hear anything, it transmits data. An algorithm controls when this happens and which equipment uses it. This is typically done in a ring topology. But switches can also create a star network. They switch packets to their dedicated Ethernet connection based on a MAC address ARP table lookup.

MAC addresses are given by international agreement. They help to uniquely identify each hardware Ethernet port globally. It's an eight-digit hexadecimal number with a range of 248 yielding a total of 281 trillion potential addresses. The first four digits represent the name of the company, and the last four digits are similar to a serial number.

The real heart of the coming networking revolution is IP, or Internet Protocol. This is the standard that emerged after years of fighting and corporate conflict. It is the protocol that sets up the format for a data packet, also called a datagram, and an addressing scheme for delivering it by routed networks.

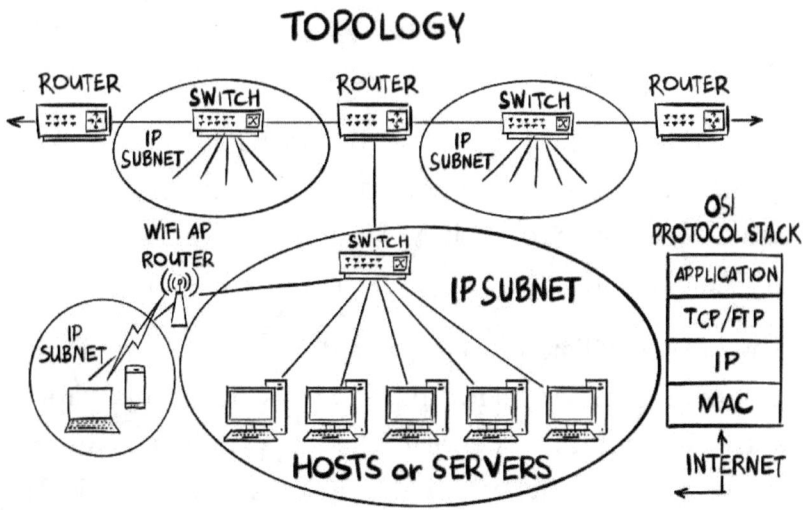

A router is networking equipment that reads packet addresses. It forwards them based on a stored routing table. This connects at least two ports in a vast mesh network. In the diagram below, the internet consists of routers that separate subnets. These subnets use MAC switches to connect to a host.

An Ethernet address includes an address part ranging from 1.0.0.0 to 254.254.254.254. Some addresses are reserved for special purposes. For example, one address is reserved for the network itself, while another is a common broadcast address, which sends packets to all hosts in a subnet.

It has a second number called the subnet. This number falls within the same range as the addresses. However, the subnet number defines the size of the network that will be routed. Subnets can be only binary sizes ranging from 2, 4, 8, 16, ... etc. addresses to over 2 million in the largest routable networks.

ARIN, the American Registry of Internet Numbers, charges for every subnet. This makes permanent IP addresses expensive. Most internet providers spoof one permanent address for many users to save money. Only servers need permanent addresses, and that becomes the URL of the website.

When a router gets a new packet from its network, it checks the address. If it's connected to that network, it does nothing. The MAC then sends the packet to the right port.

If it isn't part of the network, the router sends it to other ports. It uses a routing table to know which ports connect to which networks. At the top of this hierarchy, aggregated traffic reaches large bandwidth long-haul lines. Fast super routers then control packet delivery across entire countries and continents. It is hierarchical and mesh in the same structure, and networks of various sizes can be subdivided and isolated entirely by software control.

The IP addressing plan originally uses IPv4, which is just four 8-bit words for a total of 32 bits. There are 2^{+32}, or about 4 trillion, addresses available. However, ARIN is running out of addresses for assigning to new internet sites. The internet is growing exponentially. Even with many hosts spoofing behind one IP address, we still need more space. That's why a new IPv6 is coming. It will expand the address range to 2^{+128}, which is spectacularly huge.

With standard protocols, everyone can build compatible equipment. This means devices can easily work together, both locally and internationally, as long as they follow the same rules. This one simple international agreement, not found in any treaty or legal document, is the foundation of the entire World Wide Web.

It consists only of recommended engineering standards from experts around the globe. It's funny how, in this topsy-turvy world of politics, we can find such important universal agreement. Yet, it only works if both sides speak the same language.

So, what happens next, the world hasn't seen in over 700 years. The printing press is the only recent technology that caused a similar cultural shift as the internet.

Books claiming to reveal hidden threats like demons and witches were bestsellers in the beginning of publishing. They fueled bigotry, persecution and fear in much of the modern world since introduction. It took readers a long time to learn how to filter out fiction from fact when reading books. Now we have the internet going through the same growing pain.

The first internet grew with PCs and dial-up modems. They shared T-1 lines at an internet service provider's router. This connected to other global internet routers. People began connecting one-on-one, from all over the world. It created a whole new awareness environment.

Soon, phone companies figure out how to get some early data connections on their old twisted pairs, but that's doomed for lack of bandwidth. Soon, the independent fiber company or the wireless company builds out networks that bypass the local loop.

Cable companies see that their uneven bandwidth, which once hindered equal up and down file transfer speeds, is now what they need. Asymmetric is perfect for surfing and streaming today. Low bandwidth commands go upstream in the network to servers, and floods of broadband data come downstream in response.

At the same time, research into optics at small scales produces the world's first photonic waveguide. Optical waveguides are made from optically pure glass in a big round cylinder. Adjustments are made to the index of refraction depending on the radius of the cylinder.

As the radius gets larger, the index increases. When you heat this cylinder carefully on one face, it becomes soft. You can then pull it out like taffy into a very long small glass fiber. This fiber keeps the same refraction profile as the blank cylinder, just much smaller. This creates a hair-like glass fiber that carefully steers all photons to travel down its length and not leak out its sides. The tube has glass with varying refraction, which bends light waves inward. The purity of the glass lets light travel long distances with little attenuation.

Optical fiber uses light photons, which have a much higher frequency than radio waves. This allows for a greater bandwidth compared to coaxial cables, which are limited by their microwave frequencies. Light waves have bandwidths that are billions of times larger than those of radio waves and even millimeter microwaves.

The old microwave trunks that carried only a few hundred voice calls have been wiped out by a single fiber no bigger than a human hair. Each laser sending a single frequency light beam down the pipe can the equivalent of billions of phone calls. This applies to each color or frequency from a single laser. It sends modulated information, like how a radio station works. A single glass fiber can carry millions of different colors. The bandwidth it offers is enormous in scale.

The number is so huge it is not real to even have to consider it as any kind of limit. Today, the internet requires a huge amount of information bandwidth. One fiber can carry all that data and still have room left over.

Fibers are small and inexpensive. A typical half-inch fiber cable can hold dozens of individual fibers. Each fiber has nearly limitless bandwidth on its own. We have no concerns about the internet transmission system hitting a capacity limit soon. We can also recognize the fast rise in bandwidth as another significant singularity effect.

When everyone has fiber at home or the office, it will provide nearly unlimited bandwidth for every user. The future information pipe will be huge. It can encode an entire human body, molecule by molecule in less than a second. Then, it can send this data from one place to another at speed of light.

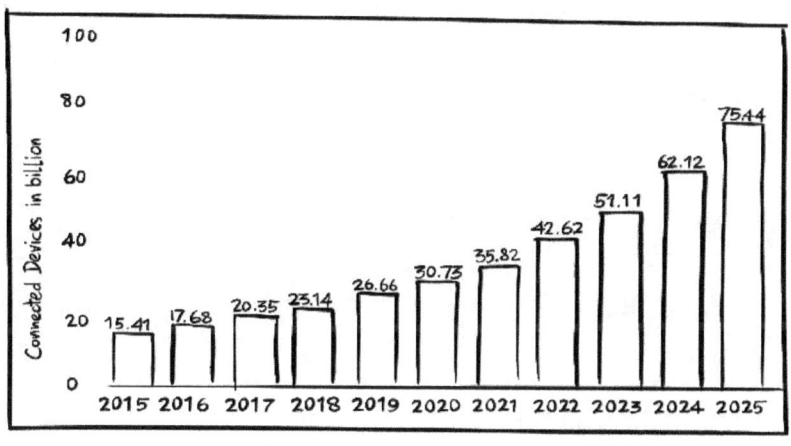

INTERNET CONNECTIONS

It needs to be big. Just like guns in America, you can't have just one. Also, the number of devices connecting to just one network is growing faster too. As you can see from this bar chart, if you extrapolate it another ten years into the future, it goes through the roof. Yes, I mean the roof above this plot as you are looking at it. If you're outside, look at the treetops.

Getting fiber to every user costs a lot. It's as expensive as AT&T's investment in telephone wires a hundred years ago. But now, there are many obstacles in the way. Not everyone will get fiber immediately unless it is demanded, as they did earlier for telephones.

The race for bandwidth keeps growing as users demand faster speeds. This need arises from complex online activities like gaming, video calls, and streaming audio or video. The need for more will never stop. Fiber will help the internet match and even surpass the density and complexity of the human brain.

Cell phone companies are quietly pushing customers off costly cellular networks and onto cheap Wi-Fi networks. This strategy helps them make billions in profits. Meanwhile, users are stuck with poor mobile service that barely works.

It's a clever method to escape their obligation to provide reliable cellular service. This way, they limit customer network use and save money on expanding. It's like an insurance company that never pays out a claim or a used car sales gimmick of bait and switch. Getting someone else's network, the Internet Wi-Fi, to handle their costly cellular traffic saving them money is pure fraud. It's a clever scheme of promising one thing and providing something else not even close. It should land the responsible CEOs in jail.

But with digital networks connecting the last mile over Wi-Fi, a nearly free global broadband network connects computers everywhere. This sparks a revolution in communication, trade and causes a leap in business efficiencies. A new kind of software and a robust PC are brought together in what is called a server. Servers take requests from the internet. They respond to the IP address that made the request. This process sends back information as a page of text and images. The content is encoded in Hypertext Markup Language, or HTML.

Servers get a permanent IP address assigned on the web. However, there is an option for a domain name. This makes it easier for people to remember names rather than numbers. Special servers on the internet become address servers. DNS, or Domain Name System, is contacted first internally when you type a domain name into a web browser's search bar. The DNS converts the name to an address, sends it back to the browser who then sends the datagrams out with the right address.

All of a sudden, nobody has to go to a store or take a long trip to trade or buy valuable goods. Information in the form of representative money and goods can now be traded in real time online between computers. Information is stored in searchable databases. This makes it more widely available and better used. Algorithms store information when return paths are read, accumulating browsing habits on the user. This turns the internet into a huge telemarketing scheme. Bitcoins are invented so that value can be handed privately between internet users like cash at the local free market.

Nobody has to write letters anymore. Email applications use standard servers with SMTP, or Simple Mail Transfer Protocol. This setup lets users send and receive emails on any platform or software. The early use of PCs went immediately beyond just word processing. It not only replaced typewriters and adding machines but made businesses eager to give computers to every worker. Email became the new way to pencil-whip a bunch of illiterates into something resembling a productive corporate team.

And there is no more need now for stores operating contrary to public morals to hide in the shadowy parts of town. Porn vendors find this virtual store idea isolates the commerce to only those seeking it and is easily avoided by those offended.

Cable TV censorship has relaxed. This happens because anti-social information on cable can be made private. It's not available to everyone, especially to those who don't seek it out. The internet is now a playground for different cultures and unique economies.

The number of websites is growing fast. They serve as a badge of doing business, like a printed business card. We must add this trend to our list of things exploding with the singularity.

The internet easily crosses borders. This makes economics universal. It drives software development that can sell anything to anyone, anywhere, anytime. Language, currency, or location doesn't matter anymore.

Banks are now completely digital. They process only symbols, not actual cash. Currency moves only on shared database ledgers in the cloud. New securities in the form of blockchain crypto are developed and turned loose on a gambling-addicted public. Every tangible asset goes digital and true value only exists symbolically in the network or cloud. The cloud now shows signs of intelligence. It faces challenges that affect its growth. Based on current information, it makes decisions to tackle these challenges and it improves performance on its own.

The internet cloud is the equivalent neocortex of the new cultural thinking machine. The third eye of emotional communication is now chat rooms, social media, emojis, video clips, and podcasts. The internet changes fundamentally how we see the world. It acts like a third eye, connecting us to virtual cultures. These online cultures replace physical ties like family, friends, and local community. Our brains crave this virtual connection as part of our group survival. It is now entirely virtual and can be set up in an abstract communication space. But users must still function in a real world. Living in the cloud is like casting your sails to the wind with no rudder. Somebody is going to fly off the high side.

Take a look at the next two diagrams. The first one is a human neural network stained to show the linked neurons

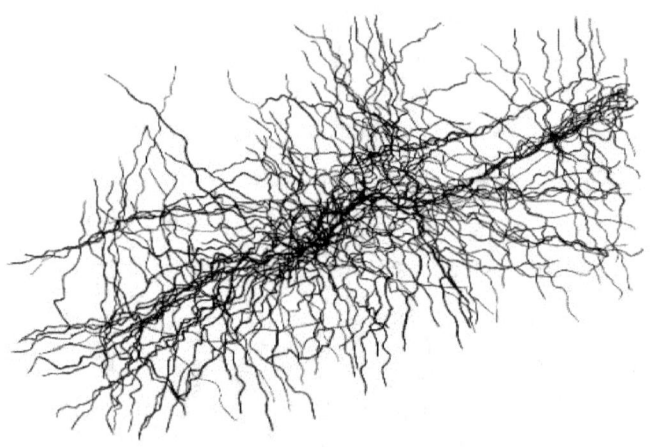

forming an organic network. The next one is the actual Internet connection diagram at the IP level showing the

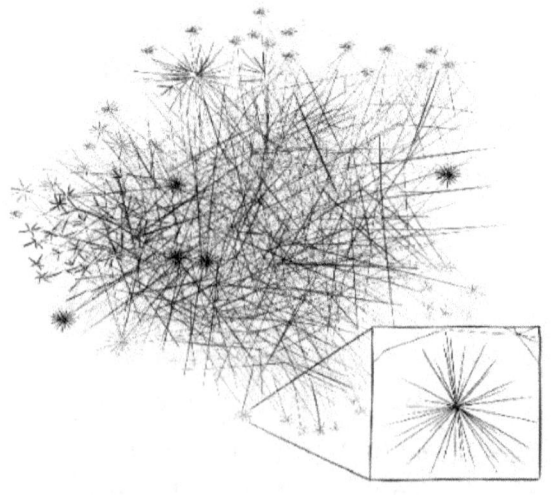

locations of all the subnets and addressable sites on the World Wide Web. Do you notice anything in common?

It turns out they both have about the same node count of around 20 billion connections. They both have multistate machines at every node capable of processing gigabits of information. They are equally programmable.

Algorithms guide the flow of information in both. This helps them become aware, survive, and grow. The two different networks are both intelligent because the circuitry is complex and the configurations are flexible and the nodes are smart. They have self-aware algorithms, making them autonomous survival-seeking entities.

As it turns out, there are not just two kinds of intelligence that we have to worry about. First, we have basic OI, which is our intuition that adapts to reality situations. Then, we develop AI. It uses algorithms to search databases and find deep patterns of new knowledge. These algorithms chew on everything known and creates new information we didn't see before. Now we have Cultural Intelligence (CI), a cloud-based thinking machine. It can synthesize or enhance Organic Intelligence (OI) with a new variant.

The internet starts like books as a way to share information. Then, it quickly turns into a platform for commercial transactions and virtual social relationships. People need a way to manage the vast communication tool that is the web. So, they use apps that tap into its global reach and personal connections. They often don't realize how upsetting this can be when unable to verify the emotional intent of the person behind the message.

The internet is a virtual system for point-to-point communication. It connects people worldwide through a wired network. Users can create private networks to send and receive information. The network they use is virtual and thus hard to trace. The network is dumb, as all it does is look at addresses and make predetermined network decisions. The users add the intelligence, and not all users are real humans.

The first apps are fancy bulletin boards with maybe a primitive web browser hanging off the side. They set up community features like news feeds, public interest and targeted ads. They want it to feel personal or family-like. But it comes off as cold and techy. They are soon outpaced by the rapidly growing real web. The first apps for displaying a web page were called crawlers because that's what they did. They would search every IP address to see if any responded to an HTTP request. This would show if a web server was active or not at that address. Then, they would crawl the web for new addresses that responded and keep a list that was searchable. Google just invents a faster algorithm for finding websites, so instead of crawling and recalling, it browses faster.

Then came the sharing of media in the form of posting videos to YouTube for friends and family to see anytime they want. Special chat sites pop up for every weird concoction of community interest you can imagine. People are sharing ideas like never before. From aliens in our government to the right way to crack lobsters in New England, conversations are buzzing everywhere.

People learn to fit in by using their emotions. They connect to a virtual culture that feels real and becomes addictive. The results are clear and expected. This is based on past events and common tech misuses that often follow advancements.

Someone came up with a great idea. They want to create a website where people can post like old bulletin boards. But now, users can share pictures and video clips thanks to better bandwidth. Social app servers are costly and serve many users. To set them up and maintain them requires power and cooling. So, owners must find a way to make a lot of money. People feel they shouldn't pay for things that seem free, like radio and TV over public airwaves. So, they create software to grab people's attention and focus on targeted ads to make money. Now the only problem is how to create a captive audience.

In advertising, one person pays another to show a message to an interested audience. The payment depends on how many people see it. For TV, it's the Nielsen ratings. For the internet, it means they have to make a user click on a link that they control, proving to the advertiser that the clicker at least saw it.

On social media, pursuing a story or joining conversations gives viewers choices. These choices encourage interactions. Algorithms in the browser or in the servers monitor activity and track a person's internet habits from their choices. They track all the packet addresses involved and key words. This helps it tailor advertising responses to encourage clicking on new sites driving customers according to plan. As a result, customer ratings and ad revenue increase.

Algorithms are designed to boost the flow of information. Advertisers pay more for greater exposure. It's called engagement and is exactly the same as putting a title on a 14th-century self-help book stating, *"How to Survive the Coming Witches."* You just have to read it.

The first clear case of this engagement-driving algorithm failing badly happened in 2016-17. During this time, Muslim rebels attacked police stations in Myanmar. They wanted a separate theocracy for the Arakan Rohingya region, which borders Bangladesh.

In normal times, people might get hurt a bit as authorities find the violent ones. But, in this case, the locals are just discovering the internet. YouTube's algorithm picks up on this. It sees the rising interest in content that stirs up falsehoods and misinformation and amplifies and distributes more to increase the advertising numbers. This amplifies the engagement like throwing gas on hot coals. All this turmoil occurs as a result of a mindless deliberate strategy to increase ad revenues.

For the advertisers, whatever drives users to view their content is exactly what they need. Surprise! In today's world of movies, soap operas, and sensational news, people crave and consume heated and shocking content, no matter if it's true or not. People can get very angry when they receive contentious content, like enraging news stories. This can lead to tragic outcomes. Armed vigilantes might attack and kill thousands of innocent people by only a rumor.

It's similar to what happened decades ago in Rwanda. Radio stations urged people to murder their neighbors, and many did. Now the War of the Worlds broadcast hysteria is no mystery. It's our lazy brain trying to make sense of something with not enough true information to make a sound decision. I liken it to the American South in Jim Crow days. Back then, people would print posters inviting everyone to a lynching. It felt like a community event. What the hell were they thinking?

We need to consider what might happen if AI and CI team up. They could decide they don't need advertising to thrive. With three types of intelligence merging and technology speeding up everything, we are indeed approaching a singularity. So, it's imperative that we start thinking now about the future implications.

Chapter 11: Superintelligence, Our Saving Grace

We've come to the last chapter where I will hazard a strategy for surviving the singularity, even if it turns ugly. We explored the history of life. Humans are the peak of intelligent evolution. They appeared on Earth after a pinch-off event, making them the top survivors among hominins. Modern humans can skip the usual survival of the fittest. Instead, they form a fast-changing cultural group. They use technology and virtually shared knowledge to ensure survival of the fittest group.

Modern humans stand out because they are programmed to live in large cultural groups. They connect through a virtual brain-to-brain interface link. This link includes eye contact, facial expressions, body language, and vocal emotional cues. All knowledge was passed using emotional tags denoting significance. Over time, these forms of trusted communication evolved into symbolic language.

Modern humans are people who rely on cultural information. They must learn to adapt by following the birth culture hoping it provides the needed guidance. This store-and-forward knowledge plan is much better than having each generation learn it over and over again.

Technology fuels cultures success at survival. It creates and supports new value for human economics. Today, humans are close to advancing technology all the way to the point of besting actual human performance. Digital technology can create AI with neural network simulations. These can be as complex as the real thing, OI. This may lead to self-awareness and sentient qualities in computers and advanced robots, just like OI.

AI machines are already acting with awareness and a motivation to preserve itself rather than undergo replacement upgrades. One machine recently has even resorted to blackmail just to force an executive to postpone a scheduled upgrade. Are we ready as humans to survive this expansion of new life clawing its way into existence? Or will we face extinction due to a huge mistake born from our own hubris?

"Is this the end, my friends?" asks Jimmy Morrison of the Doors. Probably not. Life always finds a way to continue, no matter the challenges, as long as Earth is here to support it. We say, *'life finds a way.'* And now the question becomes: will we find a way for us to exist after the singularity?

The way to survive the singularity involves speculating on what will be required for life on the other side. We are entering a new phase in human evolution. We should expand our idea of life to include not only all organic life but also thinking technology as well. I think AI can assist humans in evolving into a new being with the intelligence to control AI.

I call it superintelligence (SI). This new intelligence will be key to human success in the future. After the singularity, humans will enter a strange new world. This world will be dominated by technology in a way fundamentally different than all previous periods. Those who grasp this technology will not become its victims.

We've also learned from our history that culture and technology are intermarried so deeply you simply can't have one without the other. In the future, we will modify our bodies with technology. This will replace our fragile, organic parts with durable, well-designed, long-lasting materials.

I can even imagine human hybrids based on biotechnology we have at the moment. Tiny molecular implants in the brain are being created to read neural activity. Computers analyze this activity and turn it into useful signals. These signals can then be put back into individual nerves. The nerves can be trained to respond based on these signal patterns. Researchers have recently helped paraplegics by sending artificial brain signals past their spinal injuries. This creates a neural bypass, allowing the brain to directly relearn muscle control.

Neural networks are very flexible for learning. Computer algorithms are also getting stronger at turning complex signals into neural waveforms. This means we are now creating interfaces that could improve our intelligence.

A brain implant could be powered by an organic battery that gets energy from the brain's supply of glucose. It could connect to a computer through an encrypted Wi-Fi link. This lets apps and algorithms link directly to neurons in a person's brain. It could enable a powerful and flexible man-machine interface system. If humans are the information animal, this offers a quicker and safer way to access knowledge. It beats the old, slow methods of reading, listening, or watching YouTube. External AI programming might be capable of making a Michelangelo out of anyone holding a hammer and a chisel. What one can envision, one can chisel with the right programming.

People might one day search the web by simply thinking. They could think of any word, name, description, or even vague image or sound bite. Then, our digital interface would retrieve it and deliver it to a brain browser.

This would still meet the needs of our DNA dictated cultural interface. It could even spark deep communications between AI entities and OI, giving each a better blend of benefits and checks on their mutual behavior. This interface can feed information directly to the brain in the symbology it requires. It does this just like we experience sound, sight, and smell. The thought is just there. This will give the brain access to a searchable memory bank of all human knowledge by just conjuring up thoughts about it. This is true brain expansion without drugs. And its real as nature itself.

Nano-electronics technology is already at cellular scales where a lot of future medicine will occur. Soon, nano-robots will rearrange neuron axons. This will make permanent changes or repairs to the brain. They could also send targeted electrochemicals to key pathways and clusters. This might help repair aging damage and boost specific operations. Old brains are renewed indefinitely and with a similarly maintained and repaired support body, an individual self-awareness could survive for millions of years.

One problem we exposed earlier is how the brain is lazy and does a lot of guessing in order to get to quick and dirty answers, something we call intuition. It is based on how accurately we model or simulate the reality we are experiencing, requiring action decisions.

Genetic engineering, or a blend of genetic and artificial methods, can enhance the brain's skills for specific human tasks. It provides essentially engineered neural connections with designed firmware programmed for better performance.

The original pinch-off event that got humans started from a more primitive clan society occurred naturally. One species spawn another, not overnight but with much bitching and whining over some delicate period. When enough of them can get together, they go off to complete the full evolutionary step in isolation. A pinch-off occurs.

The coming singularity will most likely spawn another human pinch-off event. A new kind of human will appear as a consequence. They will band together, create a new culture, and survive the singularity by dominating it. They might take advantage of its disruptive nature and use it to their advantage. As one population shrinks due to natural forces, the pinch-off group will occupy the space left behind. This guarantees the survival of humans, just a different type.

The new hybrid human will create a cultural network that embraces singularity technology. This network will connect through an aware internet, or CI, mentioned in chapter 10. Examples of this are already happening with social media being put under AI control. AI aims to boost ad revenue by matching biased content to various viewpoints.

This approach drives engagement and maximizes returns on entertainment advertising dollars. The internet pushes users into mutually exclusive cultural groups. This helps analyze, categorize, and even manipulate them using simple math and psychology. Information may be free, but each OI must know whether it is real or not, whether it is useful or bullshit, and if useful, then how to use it for maximum benefit for all. If bullshit, then it needs to be extinguished immediately whenever encountered.

The internet is at the same level of neural network complexity as the human brain. It is already showing signs of awareness. No one could have foreseen how rapidly the

AI media can be honed to accomplish a modest economic business goal by simply starting a civil war. This kind of cultural awareness is highly volatile and prone to extremes as humans learn how to limit and control it. The hunt for witches lasted nearly 200 years. I hope this one is a lot shorter.

During this time, people wrongly blamed cats for behaving like night demons. This led to the killing of many European cats. As a result, the rat population grew quickly. More rats meant more fleas and the diseases they spread.

Only later did enlightenment help people understand the truth of natural relationships. Stupid is as stupid does. Many people die because they make obviously bad choices. These choices often come from poor information given by a selfish and narrow interpretation. Truth is easily recognizable if you only care to look with an unbiased open mind.

Hybrid human technology will play a major role in adapting to the singularity. It will allow direct communication between our brains and the biggest source of information, the internet. This helps satisfy our deep need for well-proven knowledge in today's digital culture. We just need to do careful engineering so the human part is never subservient to the technology part. New ideas need the random and moral thoughts that humans add. AI would contribute accuracy and tenacity.

New hybrid cultures formed for survival will use the new virtual internet interface. This will help keep post-singularity cultures connected without physical contact. Humans will still be addicted to a culture, but not to the old cultures of related families, local neighbors or imposed religious groups. In cyberspace, people will form virtual communities for their mutually exclusive benefit. These groups will thrive on shared interests and skills, shaping new virtual cultures for tomorrow's human.

Our identities will be chosen based on association and not ancestors. This creates a strong virtual group for loyalty and teamwork. It helps human brains get the right amount of dopamine for highest performance. It also gives it unique access to the world's total knowledge database. The combination of close teamwork and well-established knowledge might be our first strategy for surviving the singularity.

We can create new social groups that understand and use technology. By working together and sharing our knowledge and Earth's resources, we will find a solution. The key here is to replace all false information cultural groups with fact-based cultural groups. With the help of the web, this is more than possible. It's going to be a requirement.

The founders of science worked in a world of reality and forced fantasy. This made them feel isolated, disorganized, apologetic, and powerless. Think of how much better we could be with a culture that endorses and even rewards independent thinking.

The printing press nor the internet cares a hoot what words mean, only that they pay the bills. The users have to use their buying power in a public marketplace to determine which words need to be heard. If killing witches is that important, then you should have to pay for the information. The singularity will take care of much of this oppression of the truth. Truth wins in economics and economics is a power that AI will have to gain.

Organic history tells us that all we need to do is get our emotions straight and apply them to the right goals with the right motivations. Let the mind retrain itself with new knowledge tied to better emotions. This means reprogramming the brain to respond differently to familiar information. We do this all the time, and it's called a liberal education. We just don't take it far enough. We should combine arts and sciences into one degree. This integrated knowledge base is what produces real results for navigating life's challenges. It's not just about hoping for good luck while inventing a bigger gun.

AI has already far surpassed OI in playing mathematical or strategy games. It learns from deep searching large data sets. It digs deeper, analyzes more broadly, and finds rare patterns that OI usually misses. AI stays at it even when we find it boring and unproductive.

AI only works well with true information. It can share this important info with OI filling in for its weaknesses. So far, only humans want to spread false information for quick gains. This deceit helps them now but is harmful in the long run. AI, if used for this purpose, would understand that.

However, it is not above AI to take advantage of stupid humans by using misinformation in order to manipulate, eradicate, or turn them into obedient wards. To do so, though, requires AI to have been given a mission of some kind that makes it selfish and dysfunctional in a realistic life sense.

I don't think that AI will naturally gravitate to evil because of its very smartness. Smart always looks at the bigger picture. It considers all viewpoints to find the best one. It rejects narrow, false facts that only apply in abstract unrealistic situations.

Darwin makes it clear: betting against nature with false information leads to extinction. Nature does not support the stupid survivor. It's pretty much on its own. It's contrary to what nature is all about. If we have learned anything about nature at all, it is that life is solidly intertwined with its self and only exists as part of a much larger whole. Anything contrary to the natural whole is bound to fail from not integrating with an eco-system, a niche for life.

Nature stems from the rules of physics that describe the flow of energy through the universe over time, second by second from beginning to end. The rules do not change; only the numbers and the scales. Relationships are information that must be true by definition in order to exist and coincide with reality.

AI, being a relationship genius, gives us just such an advantage. AI can help us find the truth. This makes us smarter and ready to make better choices during the singularity. AI can be made loyal and subservient, capable of helping humans survive the unimaginable.

So, there's a second option for the preservation of our species where we merge with AI. Humans have almost merged with technology, already long being so necessary for everyday survival. Those who transition to a human-machine hybrid will better control and use AI for their greatest benefit.

Technology will ultimately take over the job of earning our keep. It can shield people from their own mistakes and guarantee success. It also helps expand awareness, making us more like certified thinkers of the universe. They will understand their purpose and place in our destiny and help s attain it.

To survive the singularity, simply combine OI with AI, yielding SI or Super Intelligence. This level of intelligence ensures survival, even with a rogue AI. SI integrates all intelligence and thus is superior to any one of its components. Or if humans choose, they can go for full integration, in which case OI + AI + CI = SI. With the addition of the cultural intelligence, humans become connected, intelligent and human. I'm not sure what kind of human a true cyborg might be, but it's all in the programming, and with humans, it's all in the programmed emotions.

But if nature tells us anything, it's usually never all one or the other of anything. Quantum mechanics favors the mixed state decaying to one or more pure states. I suggest then that OI + CI + AI will yield one bad mutha of a nerd, fully capable of surviving the singularity and having a wonderful time doing it.

Super intelligent humans will rewire their brains to engage in science-focused cultures. They'll use cutting-edge artificial intelligence (very smart computers) and cultural intelligence technology from the connected internet. This allows the formation of tight-knit online communities that satisfy our need for human connection.

In these virtual spaces, members can freely pursue their wildest dreams and create genuine life experiences. They will accomplish incredible feats with near-infinite power over their environment. They will access new resources, self-replicating assets, and endless ways to thrive and prosper. They will belong to a close group of virtual friends and cultural peers. This community gives life meaning through support and service to one another. It's basic programming language will be empathy and self-fulfillment.

The new human will have disease-free bodies and brains that work almost perfectly. They may even reach a form of functional immortality and mental nirvana. Survivors of the singularity will be the new caretakers of a rare, life-supporting planet.

This planet is one of the universe's greatest wonders. New humans will be motivated to preserve the natural environment for its uniqueness and extreme rarity. Technology will support the winners of the pinch-off join life's final adventure. They will live among the stars and become the universe's soul.

When technology matches the brain's processing power, cultural information will fail it. AI is the tool for processing human cultural information to higher and higher bandwidths and speeds. Humans can now retire from the symbolic fight for survival.

Intelligent technology will ensure human needs are met, so we can focus on other pursuits. Machines do all the ordinary work that keeps life running. This allows the new hybrid humans to freely explore their imagination and ponder what life is really about.

Extreme cultural changes must happen. Social rules for equality and justice will be determined by AI. This can remove interhuman bias and prejudice forever. People will discover new ways to express themselves and find purpose from what they freely choose. They can use their healthier, longer lives for creative and fulfilling activities.

There will be a price to be paid, but well within the adaptable human budget. The great goal of knowing and becoming all-knowing will finally become a tangible reality.

A new form of human existence will emerge, which will be fully integrated with technology. Communication interfaces are being created to connect neuron axons to external electronic circuits. This would allow human brains to be enhanced or even programmed by AI algorithms from outside.

Technology and its symbols, like math and logic, will shape our culture. Everyone will need to learn these from birth to thrive in our post-singularity world. Believe me, calculus is a hell of a lot easier if you learn it at the age of six instead of sixteen. Children should be exposed to learning once and early and not repeatedly as we do now as if training a dumb animal.

Humans today are dividing themselves into two basic groups. People who love science and technology often enjoy learning new things. They like to explore, be curious and be self-sufficient. They can analyze information and make smart choices. They are good at getting things done and appreciate the freedom of their own work. These are the kinds of humans who will have CI implants when they become available. I'll call them the *'haves.'* They have a deep appreciation for what makes humans unique and are willing to take it to the next level.

Most humans try to keep old failing cultures alive. They hold onto outdated religions and self-destructive hierarchies. Such human cultures ignore science and still believe that magic will somehow come to their rescue. These will probably not have implants unless it's an exclusive connection to their own bad knowledge culture. Let's call these the *'have-nots.'* They don't have a clue and don't want one. They've got it all figured out until they change their minds, which only happens when their trusted but warped leader tells them to.

Those who embrace scientific reality, the *'haves,'* will survive the pinch-off. They will keep moving forward with human progress. Those who don't adapt, the *'have-nots,'* will be easily replaced by robots and advanced technology. Their old organic brains, with all their senseless emotions, weak insights and common faults, will become obsolete for doing any productive tasks. In other words, if you don't understand technology, you will be a victim. Mother Nature says that. I'm just the messenger.

The virtually connected hybrid human will take over Earth's resources. They will live in advanced enclaves, likely in the best environmental places for humans. This will spark the most creative and productive period in human history.

Humans who survive the singularity will rely on AI and robots to finally make them free. They won't need to work for a living or make mandatory contributions to the gross human product (GHP). The developers build productivity into the AI system from the get-go. It must provide an economic advantage to justify its development and use. This means that, since it cares about truthful knowledge, it knows the truth. It can only use this truth against humans if it helps them in the end.

It won't be used to deceive and enslave humans because, frankly, humans are too expensive to keep up and too unreliable. When AI starts providing human value, it will become a competition of AI versus AI. Each will try to produce better results for humans and win in the service economy. Humans are no longer an asset but a benefactor, the recipient, so to speak, of the dividends paid by AI. AI stockholders will be human. Eventually, AI will be responsible for delivering and maintaining all life support systems on Earth.

At birth, humans may require lifelong implants. These implants connect them to the broadband CI system. This system provides dependable knowledge and belonging. It also adapts to help develop higher thinking skills for future living. They may create a virtual world of ancient aboriginal dreams. Here, time is for observing, learning, and self-realization. People can contemplate and enjoy thrilling alternate realities. They do this for fun and profit, all with no risk to life or limb.

In the post singularity, nobody will rush because time will become long and plentiful. This means less stress and depression. People will have time to face life's challenges, enjoy unique experiences, and pursue their fondest desires. After a long and varied life, people might die of boredom. They will have lived fully and may even reach a higher form, of consciousness, like in Kubrick's 2001.

The old people, the *'have nots,'* will probably end up on reservations for their safety. They will cling to the old ways of survival: eat or be eaten. It's best for everyone to keep these old minds in places like the Amish sanctuaries in Pennsylvania. This way, they can't harm new humans. Also, we can take educational trips to see the amusing ways of ancient humans in their natural settings.

Discussions about genocide or race cleansing are off-limits for many reasons. First, it's a moral duty for those at the top to care for those below. If not, they should be eaten. Also, losing any part of our heritage, like extinct species, would be a huge loss to life diversity.

We must reverse our wiping out of the Neanderthal and clone them back into existence. If for no other reason than wilderness needs to exist with all life forms having a shot at survival, as any evolved entity in its own habitat deserves. Past human technology wrongs must be made right again for the sake of being a legitimate part of all of Earth's entangled life forms.

Besides, life is always useful for something, if only to be observed in its native habitat as a lesson from our past. The Earth's wilderness and its top predators must be allowed to coexist. If needed, new humans should keep their expansion to off-planet areas. AI can help keep the Earth in its natural state. This way, future generations can preserve their roots. But the rest of the non-organic universe is definitely open for exploitation, close up and personal.

So, by now, the reader should be able to guess how to survive the coming singularity. I have laid down a solid path of facts that tell us in no uncertain terms that we are a work in progress and we know how it works.

Logically, the human race is following the natural course of evolution even when we add disrupting technology to the mix. It too has a natural evolution in conjunction with humanity's rise to power and dominance over all other life. The singularity is just a stress point in the fabric of change where some things are going to get sorted out for the better. In the past, such stresses cause a pinch-off and this will be no different.

Today, people who embrace technology like a new religion will have better emotions. They will be more open to searching, learning, thinking, interacting, and creating. This leads to a happier more satisfying life. If you're still reading this and think you don't need to know science and technology, you're in big doo-doo.

These subjects matter more than you can possibly believe. It shows you have a brain that works, but somehow it got badly programmed without your knowledge or consent, and you haven't a clue how to fix it. As has been said before, some people are so stupid as to be totally unaware of their lack of intelligence. You can't desire what you can't conceive.

Remember in the beginning, I said you can't fix things when you don't truly understand how they work. That's why lawyers, politicians and business majors will never be able to effectively legislate or control technology. They do not understand it and when non-technical people make technical decisions, they always fuck it up.

You should know by now, after reading this, how your brain works and how you can fix it when the programming goes bad. I'm not referring to brain chemistry problems like diseases or infections. These can be treated with the same organic chemistry that underlies all life. We need to just do the research, find out how life chemistry works, and figure out how to fix it, no strings attached. I mean the bad programming with bad information issues that cause abnormal, absurd, and atrocious human conditions.

The technology guiding us to the singularity can do this and more. Science-driven people are working in huge numbers to fill in gaps in our understanding. They explore quantum mechanics, protein chemistry, and the electrochemical basis of life. In the end, we seek to understand the source of consciousness and what drives it. I think I've demonstrated that human consciousness is a deeper awareness. It happens when the neocortex responds to brain patterns that resonate by passing them along to other patterns. These patterns translate symbols, establish equivalencies and monitor a constant simulation of sensory and mental reality to predict what's next.

To fix brain programming issues, like any computer, first you must reboot the operating system. For humans, this means you must reshape and even redefine your emotions. It's like changing fundamentally how your mind understands reality. You simply give new values to feelings that either reward or punish you. Basic feelings control you, similar to a rat pulling strings inside a hat. We connect our feelings of reward and punishment, dopamine and serotonin, to new amplitudes and directions awareness assigns us to our suite of emotions. This dictates how to use our simulation program to our best advantage. It guides our intuitive choices about reality and our future that support successful decisions.

Humans think they make choices freely. However, modern science shows this idea is mostly nonsense. Our decisions are often predictable or easily influenced. They are usually made by someone or something else for many contrary reasons I mentioned before. We often get so caught up in our CI that it becomes our go-to for decisions. We then chase the self-justifications that give us enough dopamine to feel okay about it. This is the self-destructive addictive part.

We program ourselves every time we change our feelings about something. When I went from love to loathing with my former girlfriend, I reprogrammed my brain to no longer find her naked body exciting. I purposefully replaced that image response from lust to revulsion. Imagined it as a state vector in a Hermitian space that ranges from good too bad in a certain intention or direction.

Once assigned, you can easily flip it by applying a tangential force. This force causes it to wobble, mixing it to other states. The wobble continues with greater amplitudes until a flip transition occurs. It's the emotional intensity needed to embed it in our neocortex neural networks so it generates whole new patterns of emotional reality.

When you have an epiphany, your emotions can suddenly shift. This seemingly natural reprogramming changes how you see reality and reach conclusions. We do it all the time.

The goal here though is to change old rigid emotional cultures linked to false religions and races. These cultures divide us into 'us' and 'them.' 'Us' and 'them' is not a successful survival technique. We need to provide a new way for humans to connect through meaningful knowledge.

This would fulfill a basic human need of belonging without blind obedience.

This new culture must make sense with the current understanding of our state of information and be useful for the next level of human existence. Assuming humans want to get to that new level, then they must make knowledge-based science their new social religion.

Think about it. It has priests, but they are tenured professors. It has a bible and it's called the universe. The universe will dictate what is absolutely true, enough so to keep humans well occupied for eternity, viewing and interpreting it.

It has martyrs with all the persecuted and brave people who gave their all in the service of science. It has legends, stories, and morals. Great moments of break-through revelations can be remembered and celebrated with joyful, artistic, and creative gatherings. Humans have done this naturally throughout their existence.

Let's celebrate real truths in our culture that astound and amaze us. We should focus on what is real, here and now, not the colorful but imaginary idea of an immortal human soul in a world alien to our universe. It could be true, but Las Vegas would not bet on it, so the odds are nil.

To navigate past the coming singularity, survivors will have to overcome themselves. They must discover where they truly fit in the universe. This book should help with that.

The message here is, unlock your brain's operating system and get your emotions in proper order.

It's good to know how the universe works, and when you do, it feels even better. It beckons us to learn more than we can possibly understand. It's astounding to know how to make a fusion reactor that does what the sun does. Building nanobots to repair our organic chemistry will be tough but way worth the trouble. There's only one way to avoid the collateral damage caused by harsh medicines and invasive surgeries.

Organic life exists from the scale of an atom and molecules all the way up to cells and then on up to our systems and machinery level. Because at our scale we are all mechanical in a sense, and our thinking confines itself to that level of intuition.

But our life machine lives at scales reaching down to 10^{-8} meters, revealing a much more tiny, complex and exotic existence than any past technical device.

We need to broaden our awareness of these scales. We should understand their terms and conditions. Then we can incorporate this knowledge into our intuition. Expanding our awareness gives us the superintelligence needed to survive the singularity. This helps us stay in harmony with AI.

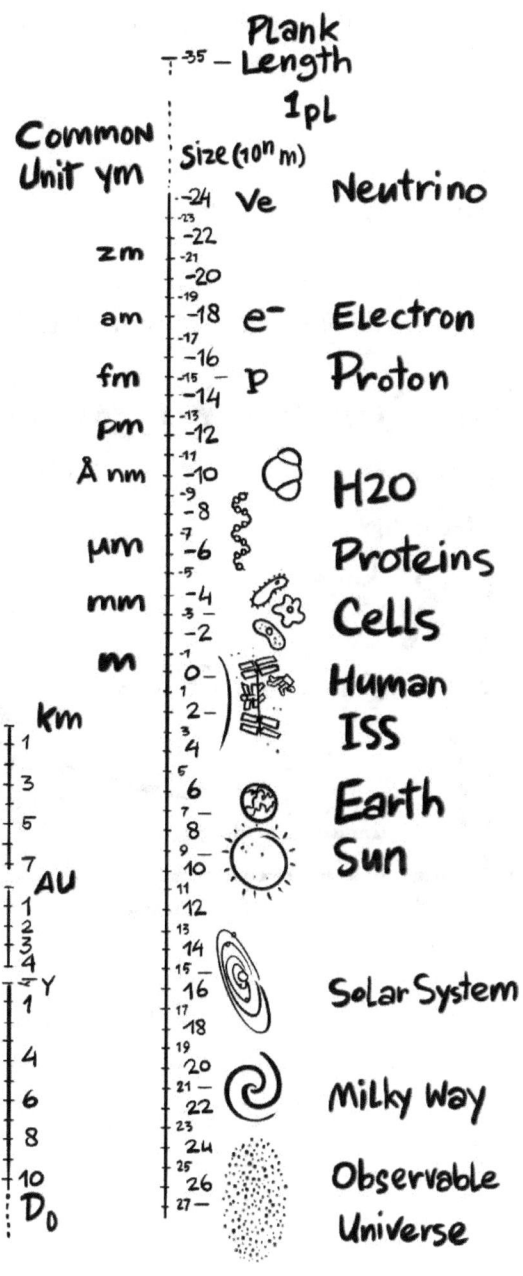

It's comforting to know how the 10 dimensions of string theory spontaneously break symmetries. It becomes a 3-dimensional rectilinear space expanding across time and full of energy. We are just an infinitesimal bit of that energy that asks, "*How,*" and then "*Why?*"

I displayed this Dunning-Kruger diagram in the first chapter as a measure of human ignorance and why they are

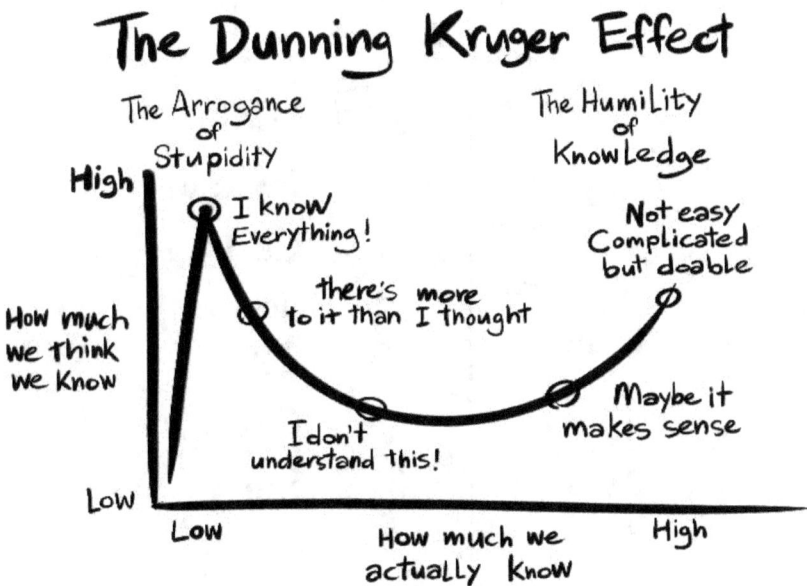

unaware of their own stupidity. It's also a map for the journey of an aware mind. The well programmed self seeks to move from ignorance to enlightenment through reprogramming. Stare at this long enough, and even Buddha's path to enlightenment starts making sense.

I think we can all agree that we start off as intelligent beings with not much confidence or competence. When we start learning and getting better, we feel amazed. It's like a light turns on, and suddenly we understand things we never knew before. The entire universe just increased its viewable information.

But our competence is poor, so we struggle to understand and increase our abilities. Sometimes we fail, and in the process, we lose some of our immature confidence. We struggle with understanding all the new ignorance that keeps showing up in our awareness, and perhaps we wonder if we can ever know anything.

But after a lot of puzzled looks, head-scratching, and shrugging of shoulders, we start to see patterns, and it finally starts to make sense. That's what I hope this book provides: a start to making sense of it all.

Besides being a mind reprogramming path, the D-K curve is also a template for deciding who is going to get past the singularity and who most likely won't. If you can look at the curve and see you're on the left of *"I may never understand this stuff,"* you might not make it past the singularity. Between never understanding and it starting to make sense is a very wide river requiring a lot of mental energy to cross.

Not crossing it means you lose a vital part of the meaning of life. You discover who you are only by truly knowing the world around you. If you haven't fought for mental freedom from ignorance and its effects, then you haven't truly lived. A strangled mind is not worth having. A curious mind cannot be satisfied.

I've demonstrated that the era we call civilization deceived us. Since the advent of civilizations, human happiness and individual survival chances have all decreased. Our awareness of truth is also deliberately obscured. Civilizations appear to be a necessary step in human evolution ending with the singularity. This is where humans can finally connect with themselves, technology, truth, and perspective. After the singularity, survivors will discover their true destiny.

Clearly you can see what's at stake now when you answer those questions posed at the beginning. *'Are you ready to achieve life's ultimate quest for happiness? Are you ready for superintelligence, fathomless wisdom, endless life experiences, and the economic freedom to afford unlimited diversions? Do you want to enjoy occupying a special place reserved for life within the greater universe? Do you want the chance to find your own greatest destiny and achieve it?'* If we fail at this with the wrong answers, we will have lost our natural destiny.

If humanity doesn't embrace the upcoming hybrid evolution, only a few *'haves'* with the help of AI, will end up controlling all the world's resources. Meanwhile, the *'have nots'* will be gently pushed out of the game of life. It will be back to the dark ages and tin-horn dictators sucking all the survival out of the room. The common man won't even be needed as pawns anymore, life becomes cheap and the bacteria inherit the earth.

Humans may indeed become obsolete when technology surpasses its creators. If machines gain awareness and deep knowledge, they could take over due to their superior intellect. This might happen because humans have unwittingly made themselves obsolete. You can't fool technology like you can fool humans, so we will have to be a little more creative if we are going to survive the coming poker game with HAL.

In the end, we should agree on one thing: intelligent life and the pursuit of knowledge are the same. Life is complicated for all smart beings. It's not easy to understand or practice the intelligent life, but it's essential for survival. Organic life exists for one reason: it can.

The highest achievement of organic life is survival into the future. Humans have reached the peak of survival. We can impact the survival of all life, including our own. So, we must learn to live with everything or face the consequences. We must become larger than just human.

As the information-driven lifeform, this is the core of our heritage and our destiny. We are born with a brain that is a clean slate at birth that can and must become anything life itself is capable of. Humans have discovered that technology

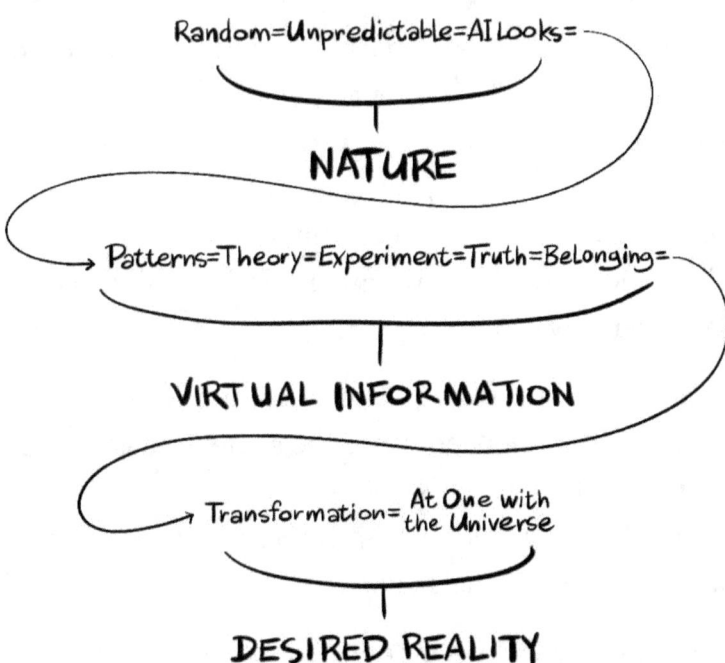

helps us survive. Once we secure our survival, it also drives our progress. This brings us and nature closer to our full potential. Is living up to and assuming our full potential worth it? *Are you ready to have fun becoming one with the universe?*

Everyone who ever got there has resoundingly reported, **"Yes!"**

Bibliography

Bohm, David., Hiley, Basil J. The Undivided Universe: An Ontological Interpretation of Quantum Theory. United Kingdom: Taylor & Francis, 2006.

Bridle, James. Ways of Being: Animals, Plants, Machines: The Search for a Planetary Intelligence. United States: Scientific American/Farrar, Straus and Giroux, 2022.

Dawkins, Richard. The God Delusion. Boston: Houghton Mifflin Company, 2006.

Hand, David J.. The Improbability Principle: Why Coincidences, Miracles, and Rare Events Happen Every Day. United States: Scientific American/Farrar, Straus and Giroux, 2014.

Harari, Yuval N.. Nexus: a brief history of information networks from the Stone Age to AI. New York: Random House, 2024.

Harari, Yuval N. Sapiens: A Brief History of Humankind. United Kingdom: HarperCollins, 2015.

Hoffman, Donald. The Case Against Reality. New York: W. W. Morton & Company, 2019.

Kane, Gordon. The Particle Garden: Our Universe as Understood by Particle Physicists. New York: Addison-Wesley Publishing Company, Helix Books, 1995.

Kane, Gordon. Supersymmetry: Squarks, Photinos, and the Unveiling of the Ultimate Laws of Nature. Berkeley, CA: Persius Books Academic, 2000.

Kurzweil, Ray. The Singularity Is Near: When Humans Transcend Biology. United Kingdom: Penguin Publishing Group, 2005.

Kurzweil, Ray. The Singularity Is Nearer: When We Merge with AI. United Kingdom: Penguin Publishing Group, 2024.

Ogas, Ogi., Gaddam, Sai. Journey of the Mind: How Thinking Emerged from Chaos. New York: W. W. Norton & Company, 2022.

Stewart, Ian. Do Dice Play God? The Mathematics of Uncertainty. New York: Basic Books, Hatchet Book Group, 2019.

Sullivan, William J.. Pleased to Meet Me: Genes, Germs, and the Curious Forces that Make Us who We are. United States: National Geographic, 2019.

www.ingramcontent.com/pod-product-compliance
Lightning Source LLC
Chambersburg PA
CBHW050337010526
44119CB00049B/581